International Conference on Heavy Ion Collisions in the LHC Era

Journal of Physics: Conference Series Volume 422

Quy Nhon, Vietnam
16-20 July 2012

ISBN: 978-1-62748-299-8
ISSN: 1742-6588

Printed from e-media with permission by:

Curran Associates, Inc.
57 Morehouse Lane
Red Hook, NY 12571

Some format issues inherent in the e-media version may also appear in this print version.

Printed by Curran Associates, Inc. (2013)

Institute of Physics
Dirac House, Temple Back
Bristol BS1 6BE UK

Phone: 44 1 17 929 7481
Fax: 44 1 17 920 0979

techtracking@iop.org

Additional copies of this publication are available from:

Curran Associates, Inc.
57 Morehouse Lane
Red Hook, NY 12571 USA
Phone: 845-758-0400
Fax: 845-758-2634
Email: curran@proceedings.com
Web: www.proceedings.com

Foreword

Papers

FOREWORD

The International Conference on Heavy Ion Collisions in the LHC Era was held in Quy Nhon, Vietnam, on 16–20 July 2012. The series Rencontres du Vietnam, created by Jean Tran Thanh Van in 1993, consists of international meetings aimed to stimulate the development of advanced research in Vietnam and more generally in South East Asia, and to establish collaborative research networks with Western scientific communities. This conference, as the whole series, also supports the International Center for Interdisciplinary Science Education being built in Quy Nhon.

The articles published in this volume present the latest results from the heavy-ion collision programs of RHIC and LHC as well as the corresponding theoretical interpretation and future perspectives. Lower energy nuclear programs were also reviewed, providing a rather complete picture of the state-of-the-art in the field.

We wish to thank the sponsors of the Conference on Heavy Ion Collisions in the LHC Era: the European Research Council; Xunta de Galicia (Spain); EMMI (Germany) and Agence Nationale de la Recherche (France) François Arleo (Laboratoire d'Annecy-le-Vieux de Physique Théorique, France)

Francois Arleo, Carlos A Salgado and Jean Tran Thanh Van

Peer review statement

Abstract Metrics

All papers published in this volume of *Journal of Physics: Conference Series* have been peer reviewed through processes administered by the proceedings Editors. Reviews were conducted by expert referees to the professional and scientific standards expected of a proceedings journal published by IOP Publishing.

Journal of Physics: Conference Series **422** (2013) 012001 doi:10.1088/1742-6596/422/1/012001

Jets, Bulk Matter, and their Interaction in Heavy Ion Collisions at the LHC

K. Werner[(a)], B. Guiot[(a)], Iu. Karpenko[(b)], M. Bleicher[(c)], T. Pierog[(d)], S. Porteboeuf-Houssais[(e)]

[(a)] SUBATECH, University of Nantes – IN2P3/CNRS– EMN, Nantes, France
[(b)] Bogolyubov Institute for Theoretical Physics, Kiev 143, 03680, Ukraine
[(c)] FIAS, Johann Wolfgang Goethe Universitaet, Frankfurt am Main, Germany
[(d)] Karlsruhe Inst. of Technology, KIT, Campus North, Inst. f. Kernphysik, Germany
[(e)] LPC, Universit Blaise Pascal, CNRS-IN2P3, Clermont-Ferrand, France

E-mail: werner@subatech.in2p3.fr

Abstract. We discuss a theoretical scheme that accounts for bulk matter, jets, and the interaction between the two. The physical picture of our approach is the following: Initial hard scatterings result in mainly longitudinal flux tubes, with transversely moving pieces carrying the p_t of the partons from hard scatterings. These flux tubes constitute eventually both bulk matter (which thermalizes, flows, and finally hadronizes) and jets, according to some criteria based on partonic energy loss. High energy flux tube segments will leave the fluid, providing jet hadrons via the usual Schwinger mechanism of flux-tube breaking caused by quark-antiquark production. But the jets may also be produced at the freeze-out surface. Here we assume that the quark-antiquark needed for the flux tube breaking is provided by the fluid, with properties (momentum, flavor) determined by the fluid rather than the Schwinger mechanism. Considering transverse fluid velocities up to 0.7c, and thermal parton momentum distributions, one may get a "push" of a couple of GeV to be added to the transverse momentum of the string segment. This will be a crucial effect for intermediate p_t jet hadrons.

The starting point is a multiple scattering approach corresponding to a marriage of Gribov-Regge theory and perturbative QCD (pQCD), for details see [1]. An elementary scattering corresponds to a parton ladder, containing a hard scattering calculable based on pQCD, including initial and final state radiation. These ladders are identified with flux tubes, which are mainly longitudinal objects, with transversely moving parts, carrying the transverse momenta of the hard scatterings. These objects are also referred to as kinky strings [2, 3, 4]. The consistent quantum mechanical treatment of the multiple scattering is quite involved, it is based on cutting rule techniques to obtain partial cross sections, which are then simulated with the help of Markov chain techniques [5].

Strings break via the production of quark-antiquark pairs according to the so-called area law [1, 5, 6, 7]. The string segments are identified with final hadrons and resonances. In heavy ion collisions and also in high multiplicity events in proton-proton scattering at very high energies, the density of strings will be so high that the strings cannot decay independently as described above. Here we have to modify the procedure as discussed briefly in the following. All the details can be found in ref. [8].

The starting point are still the flux tubes (kinky strings) originating from elementary collisions. These flux tubes will constitute both, bulk matter which thermalizes and expands collectively, and jets. The criterion which decides whether a string piece ends up as bulk or jet, will be based on energy loss. In the following we consider a flux tube in matter, where "matter" first means the presence of a high density of other flux tubes, which then thermalize.

Three possibilities should occur, referred to as A, B, C, see Fig. 1(a,b) :

A String segments far from the surface and/or being slow will simply constitute matter, they lose their character as individual strings. This matter will evolve hydrodynamically and finally hadronize ("soft hadrons").

B Some string pieces (like those close to transversely moving kinks) will be formed outside the matter, they will escape and constitute jets ("jet hadrons").

C There are finally also string pieces produced inside matter or at the surface, but having enough energy to escape and show up as jets ("jet hadrons"). They are affected by the flowing matter ("fluid-jet interaction").

The surface is defined by a constant temperature. In principle the formation and expansion of matter and the interaction of partons with matter is a dynamical process. However, the initial distribution of energy density and the knowledge of the initial momenta of partons (or string segments) allows already an estimate about the fate of the string segments. By "initial time" we mean some early proper time τ_0 which is a parameter of the model.

Only the bulk segments are used to determine the initial conditions for hydrodynamics, following the same procedure as explained in [1, 8]. Starting from this initial condition, the bulk matter will evolve according to the equations of ideal hydrodynamics till "hadronization", which occurs at some "hadronization temperature" T_H [1]. Hadronization means that we change from matter description to particle description, but hadrons still interact among each other, realized via a hadronic cascade procedure [9], already discussed in [1].

Most interesting are the segments which are formed inside but still escape. These are type C segments. They escape, but their properties change. Actually such a segment leaves "matter" at the hadronization surface at a particular space-time point x, which is characterized by some collective flow velocity $\vec{v}(x)$. We assume that the string breaking in this case is modified such that the quark and antiquark (or diquark) necessary for the string breaking are taken from the flowing fluid rather than being produced via the Schwinger mechanism. So the new string segment is composed of a quark and antiquark (diquark) carrying the flow velocity, and the string piece in between, which has not been changed. This string piece may or may not carry large momentum, depending on whether it is close to a kink or not, the former possibility shown in Fig. 1(b).

Figure 1. (Color online) Flux tube in matter (from other flux tubes, blue colored area). One distinguishes three types of behavior for string segments, noted as A, B, C (see text). The highest p_t string segment may be of type B (a) or of type C (b). A type C segment picks up quark and antiquark from the fluid, carrying momenta and flavor according to the fluid properties (b).

Journal of Physics: Conference Series **422** (2013) 012001 doi:10.1088/1742-6596/422/1/012001

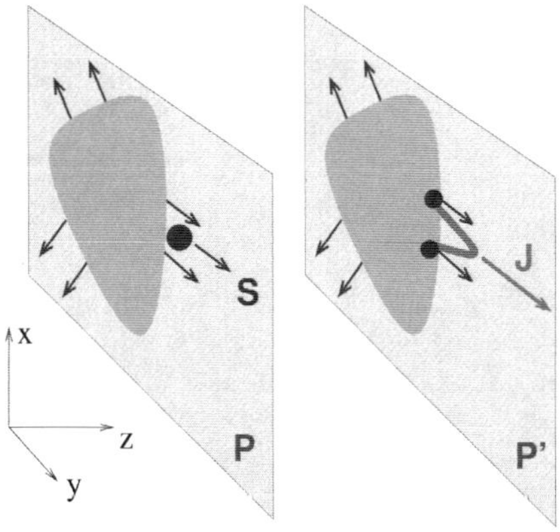

Figure 2. (Color online) Sketch of two cuts of the fluid volume corresponding to the space-time rapidities η_s and η_s', the two corresponding transverse planes being $\mathbf{P}$ and $\mathbf{P}'$. We show the example of a triangular flow pattern – the same at η_s and η_s'.

Our prescription for bulk-jet separation and interaction affects strongly dihadron correlations and azimuthal anisotropy results at intermediate p_t, due to the fluid-jet interaction. Let us consider the situation of an initial azimuthal anisotropy in the energy density which is transported into a corresponding anisotropy in the flow. We sketch in Fig. 2 the (somewhat exaggerated) situation of a triangular transverse flow pattern with maximal flow around $\phi = 0^o$, 120°, and 240° (with respect to the y-axis). The flow maxima are indicated by blue arrows. Again it is very important that this flow pattern is (not necessarily in magnitude, but in shape) very similar at different longitudinal positions – in the figure indicated by the two transverse planes $\mathbf{P}$ and $\mathbf{P}'$, corresponding to two different space-time rapidities η_s and η_s'. A soft hadron ($\mathbf{S}$) produced at η_s at the fluid surface close to the position of maximal flow (for example at $\phi = 0^o$), will be boosted by the latter one and therefore carry information about this flow. A jet hadron ($\mathbf{J}$) produced at η_s' at the same angle ($\phi = 0^o$) close to the surface, will pick up a quark and an antiquark, both carrying flow, which adds the corresponding transverse momentum to the p_t of the string segment (red element in the figure). It is the same flow which affects the jet hadron at η_s' and the soft hadron at η_s, which creates dihadron correlations. It also explains anisotropic behavior at quite large p_t. In Fig. 3, we plot v_2 as a function of the transverse momentum for different centralities in Pb-Pb collisions at 2.76 TeV. The magnitude of the elliptical flow coefficients increase at low p_t to reach a maximum around 2-3.5 GeV/c and then drop slowly at large p_t. The behavior at high p_t is the most interesting aspect: even at 10 GeV/c, there is a significant amount of elliptical flow, due to the fluid-jet interaction, which pushes jet particles in the direction of the collective flow at the freeze-out surface.

To summarize: We presented a theoretical scheme which accounts for bulk matter, jets, and the interaction between the two. The criterion for bulk-jet separation is based on parton energy loss. But in addition to the latter mechanism, there are very important new phenomena which have not been discussed so far: The interaction between jet hadrons and soft ones (from fluid freeze-out), and the interaction between the fluid and jet hadrons at the moment of the creation of the latter ones. Particle production between zero and (at least) 20 GeV/c is affected. We understand quantitatively azimuthal anisotropies in single particle production and dihadron (long range) correlations at large values of p_t.

ACKNOWLEDGMENTS: This research has been carried out within the scope of the ERG (GDRE) "Heavy ions at ultra-relativistic energies", a European Research Group comprising IN2P3/CNRS, Ecole des Mines de Nantes, Universite de Nantes, Warsaw University of

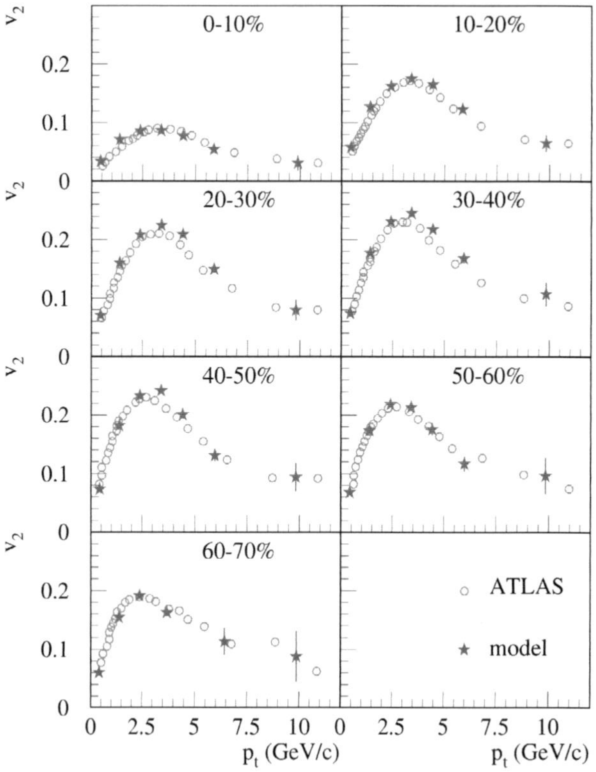

Figure 3. (Color online) p_t dependence of elliptical flow for different centralities in Pb-Pb collisions at 2.76 TeV. We compare the ATLAS data [11] (circles) with calculations (red lines).

Technology, JINR Dubna, ITEP Moscow, and Bogolyubov Institute for Theoretical Physics NAS of Ukraine. Iu.K. acknowledges partial support by the State Fund for Fundamental Researches of Ukraine (Agreement F33/42-2012) and National Academy of Sciences of Ukraine (Agreement F3/2012). . M.B. thanks the Hessian LOEWE initiative for financial support.

[1] K. Werner, Iu. Karpenko, T. Pierog, M. Bleicher, K. Mikhailov, Phys. Rev. C 82, 044904 (2010)
[2] Y. Nambu, Proc. Intl. Conf. on Symmetries and Quark Models, Wayne State Univ., 1969
[3] J. Scherk, Rev. Mod. Phys. 47, 123 (1975)
[4] C. Rebbi, Phys. Rep. 12 (1974) 1
[5] H. J. Drescher, M. Hladik, S. Ostapchenko, T. Pierog and K. Werner, Phys. Rept. 350, 93, 2001
[6] X. Artru, Phys. Rep. 97, 147 (1983)
[7] D. A. Morris, Nucl. Phys. B288, 717, 1987
[8] K.Werner, Iu.Karpenko, M. Bleicher, T.Pierog, S. Porteboeuf-Houssais, arXiv:1203.5704, Phys. Rev. C 85, 064907 (2012)
[9] M. Bleicher et al., J. Phys. G25 (1999) 1859; H. Petersen, J. Steinheimer, G. Burau, M. Bleicher and H. Stocker, Phys. Rev. C78 (2008) 044901
[10] ALICE collaboration, K. Aamod et al., Phys.Lett. B708 (2012) 249-264, arXiv:1109.2501
[11] ATLAS collaboration, G. Aad et al., Phys.Lett. B707 (2012) 330-348, arXiv:1108.6018

Journal of Physics: Conference Series **422** (2013) 012002 doi:10.1088/1742-6596/422/1/012002

Recent elliptic flow results from Beam Energy Scan at STAR

Shusu Shi (for the STAR collaboration)

1. Institute of Particle Physics, Central China Normal University, Wuhan, Hubei, 430079, China

2. The Key Laboratory of Quark and Lepton Physics (Central China Normal University), Ministry of Education, Wuhan, Hubei, 430079, China

E-mail: shishusu@gmail.com

Abstract. We present measurement of elliptic flow, v_2, for charged and identified particles at midrapidity in Au+Au collisions at $\sqrt{s_{NN}}$ = 7.7 - 39 GeV. We compare the inclusive charged hadron v_2 to those from high energies at RHIC ($\sqrt{s_{NN}}$ = 62.4 and 200 GeV), at LHC ($\sqrt{s_{NN}}$ = 2.76 TeV). The energy dependence of the difference in v_2 between particles and anti-particles is discussed.

1. Introduction

Searching for the region of a possible phase transition between the Quark Gluon Plasma (QGP) and the hadron gas phase in the QCD phase diagram is one of the main goals of the Beam Energy Scan (BES) at RHIC. Due to the sensitivity to the underlying dynamics in the early stage of the collisions, the elliptic flow (v_2) could be used as a powerful tool [1]. In the top energy ($\sqrt{s_{NN}}$ = 200 GeV) for Au+Au and Cu+Cu collisions at RHIC, the number of constituent quark (NCQ) scaling in v_2 reflects that the collectivity has been built up at the partonic stage [2, 3, 4]. Especially, the NCQ scaling of multi-strange hadrons, ϕ and Ω, provides the clear evidence of partonic collectivity because they are less sensitive to the late hadronic interactions [5, 6]. Further, a study based on a multi-phase transport model (AMPT) indicates the NCQ scaling is related to the degrees of freedom in the system [7]. The holding of the NCQ scaling reflects the partonic degree of freedom, whereas the breaking of the scaling reflects the hadronic degree of freedom. In reference [8], the importance of ϕ meson has been emphasized. Without partonic phase, the ϕ meson v_2 could be small or zero. Thus, the measurements of elliptic flow with the BES data offer us the opportunity to investigate the phase boundary in the QCD phase diagram.

In this proceedings, we present the v_2 results of charged and identified hadrons from the STAR experiment in Au+Au collisions at $\sqrt{s_{NN}}$ = 7.7 - 39 GeV. The Time Projection Chamber (TPC) [9] is used as the main detector for event plane determination. The centrality is determined by the number of tracks from the pseudorapidity region $|\eta| \leq 0.5$. The particle identification for $\pi^{\pm}$, $K^{\pm}$ and p ($\overline{p}$) is achieved via the energy loss in the TPC and the time of flight information from the multi-gap resistive plate chamber detector [10]. Strange hadrons are reconstructed from their decay channels: $K_S^0 \rightarrow \pi^+ + \pi^-$, $\phi \rightarrow K^+ + K^-$, $\Lambda \rightarrow p + \pi^-$ ($\overline{\Lambda} \rightarrow \overline{p} + \pi^+$), and $\Xi^- \rightarrow \Lambda + \pi^-$ ($\overline{\Xi}^+ \rightarrow \overline{\Lambda} + \pi^+$). The detailed description of

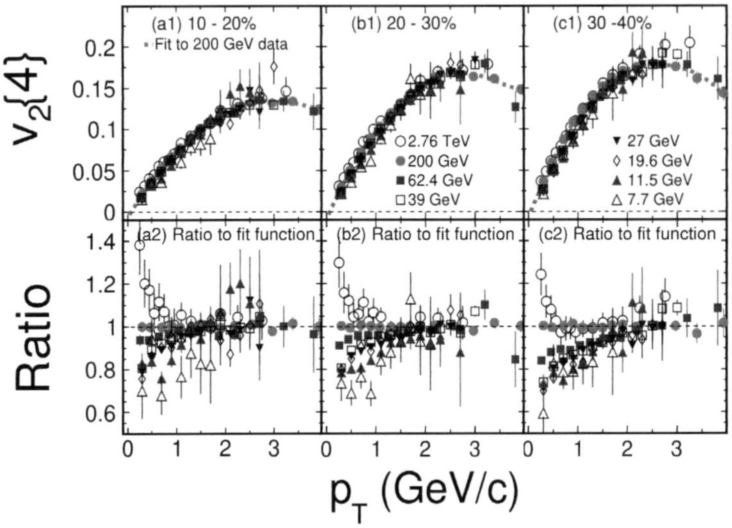

Figure 1. The top panels show $v_2\{4\}$ vs. p_T at midrapidity for various collision energies ($\sqrt{s_{NN}}$ = 7.7 GeV to 2.76 TeV). The results for $\sqrt{s_{NN}}$ = 7.7 to 200 GeV are for Au+Au collisions [15] and those for 2.76 TeV are for Pb + Pb collisions [16]. The dashed red curves show the fifth order polynomial fits to the results from Au+Au collisions at $\sqrt{s_{NN}}$ = 200 GeV. The bottom panels show the ratio of $v_2\{4\}$ vs. p_T for all $\sqrt{s_{NN}}$ with respect to the fit curve. The results are shown for three collision centrality classes: 10 – 20% (a1), 20 – 30% (b1) and 30 – 40% (c1) [17].

the reconstruction procedure can be found in Refs. [2, 3, 11]. The event plane method [12] and cumulant method [13, 14] are used for the v_2 measurement.

2. Results and Discussions

Figure 1 [17] shows the p_T dependence of $v_2\{4\}$ from $\sqrt{s_{NN}}$ = 7.7 GeV to 2.76 TeV in 10 – 20% (a1), 20 – 30% (b1) and 30 – 40% (c1) centrality bins, where the ALICE results in Pb + Pb collisions at $\sqrt{s_{NN}}$ = 2.76 TeV are taken from Ref. [16]. The 200 GeV data is empirically fit by a fifth order polynomial function. For comparison, the v_2 from other energies are divided by the fit and shown in the lower panels of Fig. 1. We chose 200 GeV data as the reference, because the statistical errors are smallest. For p_T below 2 GeV/c, the v_2 values increase with collision energy. Beyond p_T = 2 GeV/c the v_2 results are comparable within statistical errors. The increase of $v_2(p_T)$ at low p_T as a function of energy could be due to the change of chemical composition from low to high energies and/or larger collectivity at higher collision energy.

Figure 2 [18, 19] shows the excitation function for the relative difference of v_2 between particles and anti-particles. In order to reduce the non-flow effect, the η-sub event plane method is used for the measurement. The η-sub event plane method is similar to the event plane method, except one defines the flow vector for each particle based on particles measured in the opposite hemisphere in pseudorapidity. An η gap of $|\eta| < 0.05$ is used between negative/positive η sub-event to guarantee that non-flow effects are reduced by enlarging the separation between the correlated particles. The difference between the v_2 values of the baryon and the corresponding anti-baryon is within 10% at $\sqrt{s_{NN}}$ = 39 and 62.4 GeV, while the difference increases as the beam energy decreases to below 39 GeV. At $\sqrt{s_{NN}}$ = 7.7 GeV, the difference in the v_2 values of the protons versus anti-protons is around 60%. The difference between v_2 values for the π^+

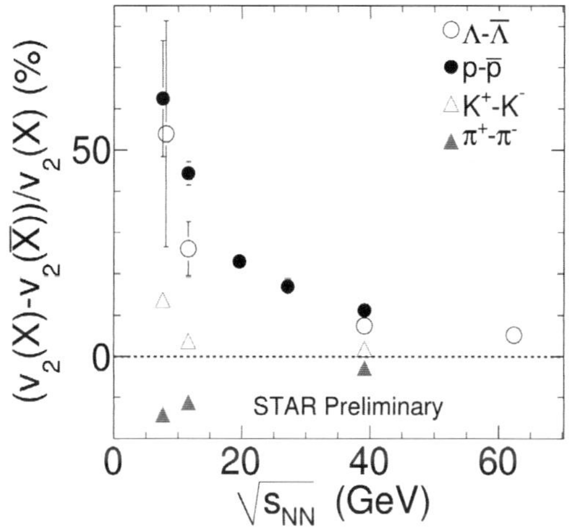

Figure 2. (Color online) The difference of v_2 for particles and anti-particles ($v_2(X) - v_2(\overline{X})$) divided by particle v_2 ($v_2(X)$) as a function of beam energy in Au+Au collisions (0-80%) [18, 19].

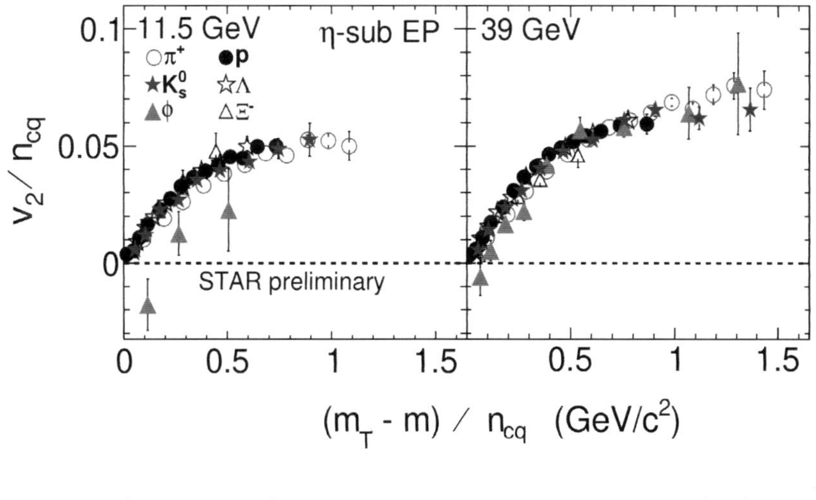

Figure 3. (Color online) The number of constituent quark ($n_{\rm cq}$) scaled v_2 as a function of transverse kinetic energy over $n_{\rm cq}$ ($(m_T - m)/n_{\rm cq}$) for various identified particles in Au+Au collisions (0-80%) at $\sqrt{s_{NN}}$ = 11.5 and 39 GeV [22, 23].

versus π^- is within 3% and those for the K^+ versus K^- is within 2% at $\sqrt{s_{NN}}$ = 39 GeV. With the decrease of beam energy, the v_2 values of π^+ versus π^- and K^+ versus K^- start to show the difference. The v_2 of π^- is larger than that of π^+ and the v_2 of K^+ is larger than that of K^-. This difference between particles and anti-particles could be qualitatively reproduced by considering the baryon transport effects [20] or by introducin hadronic potential effect in model calculations [21]. These results indicate that the hadronic interaction become more dominant at the lower beam energy. The immediate consequence of the significant difference between baryon and anti-baryon v_2 is that the NCQ scaling is broken between particles and anti-particles when $\sqrt{s_{NN}} <$ 39 GeV. Figure 3 [22, 23] shows the $m_T - m$ differential v_2 for the selected identified particles. The v_2 and $m_T - m$ ($m_T = \sqrt{p_T^2 + m^2}$) are divided by number of constituent quark in each hadron. The m represents the invariant mass for the corresponding particle. The similar scaling behavior at $\sqrt{s_{NN}}$ =39 GeV is observed as was found in Au+Au collisions at $\sqrt{s_{NN}}$ = 200 GeV. Especially, the v_2 of the ϕ mesons which are less sensitive to the later hadronic interactions follow the same trend of v_2 as for the other particles. It suggests that the partonic degree of freedom and collectivity has been built up at $\sqrt{s_{NN}}$ = 39 GeV. Whereas, at $\sqrt{s_{NN}}$ = 11.5 GeV, the v_2 for ϕ mesons falls off from other particles. The mean deviation from the v_2 values of π^+ is 2.6 σ. More events at 11.5 GeV are needed for a clear conclusion.

3. Summary

In summary, we present the v_2 measurements for charged hadrons and identified hadrons in Au+Au collisions at $\sqrt{s_{NN}}$ = 7.7 - 39 GeV. The comparison with Au+Au collisions at higher energies at RHIC ($\sqrt{s_{NN}}$ = 62.4 and 200 GeV) and at LHC (Pb + Pb collisions at $\sqrt{s_{NN}}$ = 2.76 TeV) shows the $v_2\{4\}$ values at low p_T ($p_T <$ 2.0 GeV/c) increase with increase in the collision energy. The difference between the v_2 of particles and anti-particles is observed. The baryon and anti-baryon v_2 show significant difference for $\sqrt{s_{NN}} <$ 39 GeV. The difference of v_2 between difference particles and anti-particles (pions, kaons, protons and Λs) increases with decreasing of the beam energy. Experimental data indicates that the hadronic interactions become more important at the lower beam energy.

4. Acknowledgments

This work was supported in part by the National Natural Science Foundation of China under grant No. 11105060, 10775060 and 11135011.

[1] S. A. Voloshin, A. M. Poskanzer and R. Snellings, arXiv:0809.2949.
[2] J. Adams *et al.* (STAR Collaboration), Phys. Rev. Lett. **92**, 052302 (2004).
[3] J. Adams *et al.* (STAR Collaboration), Phys. Rev. Lett. **95**, 122301 (2005).
[4] B. I. Abelev *et al.*, (STAR Collaboration), Phys. Rev. **C 81**, 044902 (2010).
[5] B. I. Abelev *et al.*, (STAR Collaboration), Phys. Rev. Lett. **99** 112301 (2007).
[6] S. S. Shi (for the STAR collaboration), Nucl. Phys. A **830**, 187c (2009); Nucl. Phys. A **862-863**, 263 (2011).
[7] F. Liu, K.J. Wu, and N. Xu, J. Phys. G **37** 094029(2010).
[8] B. Mohanty and N. Xu, J. Phys. G **36**, 064022(2009).
[9] K. H. Ackermann *et al.* (STAR Collaboration), Nucl. Instrum. Methods A **499**, 624 (2003).
[10] W. J. Llope (STAR TOF Group), Nucl. Instr. and Meth. B **241**, 306 (2005).
[11] C. Adler *et al.* (STAR Collaboration), Phys. Rev. Lett. **89**, 132301 (2002).
[12] A. M. Poskanzer and S. A. Voloshin, Phys. Rev. C **58** 1671 (1998).
[13] N. Borghini, P. M. Dinh, and J.-Y. Ollitrault, Phys. Rev. C **63**, 054906 (2001).
[14] A. Bilandzic, R. Snellings and S.Voloshin, Phys. Rev. C **83**, 044913 (2011).
[15] Y. Bai, Ph.D. thesis, Nikhef and Utrecht University, 2007.
[16] K. Aamodt *et al.* (ALICE Collaboration), Phys. Rev. Lett. **105**, 252302 (2010).
[17] L. Adamczyk *et al.*, (STAR Collaboration), arXiv:1206.5528.
[18] A. Schmah (for the STAR collaboration), arXiv:1202.2389.
[19] S. S. Shi (for the STAR collaboration), arXiv:1201.3959.
[20] J. Dunlop, M.A. Lisa and P. Sorensen, Phys. Rev. C **84**, 044914 (2011).
[21] J. Xu *et al.*, Phys. Rev. C **85**, 041901 (2012).
[22] A. Schmah (for the STAR collaboration), J. Phys. G **38** 124049 (2011).
[23] S. S. Shi (for the STAR collaboration), Acta Physica Polonica B Proceedings Supplement **5** 311 (2012).

Journal of Physics: Conference Series **422** (2013) 012003 doi:10.1088/1742-6596/422/1/012003

Boltzmann equation with a nonlocal collision term and the resultant dissipative fluid dynamics

Amaresh Jaiswal, Rajeev S Bhalerao and Subrata Pal

Tata Institute of Fundamental Research, Homi Bhabha Road, Mumbai 400005, India

E-mail: amaresh@tifr.res.in, bhalerao@tifr.res.in, spal@tifr.res.in

Abstract. Starting with the relativistic Boltzmann equation where the collision term was generalized to include gradients of the phase-space distribution function, we recently presented a new derivation of the equations for the relativistic dissipative fluid dynamics. We compared them with the corresponding equations obtained in the standard Israel-Stewart and related approaches. Our method generates all the second-order terms that are allowed by symmetry, some of which have been missed by the traditional approaches, and the coefficients of other terms are altered. The first-order or Navier-Stokes equation too receives a small correction. Here we outline this work for the general audience.

1. Introduction

The kinetic or transport theory of gases is a microscopic description in the sense that detailed knowledge of the motion of the constituents is required. Fluid dynamics (also sloppily called hydrodynamics) is an effective (macroscopic) theory that describes the slow, long-wavelength motion of a fluid close to local thermal equilibrium. No knowledge of the motion of the constituents is required to describe observable phenomena. Quantitatively, if l denotes the mean free path, τ the mean free time, k the wave number, and ω the frequency, then $kl \ll 1$, $\omega\tau \ll 1$ is the hydro regime, $kl \simeq 1$, $\omega\tau \simeq 1$ the kinetic regime, and $kl \gg 1$, $\omega\tau \gg 1$ the free-particle regime.

Hydrodynamic equations are a set of coupled partial differential equations for number density n, energy density ϵ, pressure P, hydrodynamic four-velocity u^μ, and dissipative fluxes such as bulk viscosity Π, heat current n^μ, and shear stress tensor $\pi^{\mu\nu}$. In addition, the equation of state (EoS) needs to be supplied. Hydrodynamics is a powerful technique: Given the initial conditions and the EoS, it predicts the evolution of the matter. Its limitation is that it is applicable at or near (local) thermal equilibrium only.

Hydrodynamics finds applications in cosmology, astrophysics, high-energy nuclear physics, etc. In relativistic heavy-ion collisions, it is used to calculate the multiplicity and transverse momentum spectra of hadrons, anisotropic flows and femtoscopic radii. Energy density or temperature profiles resulting from the hydrodynamic evolution are needed in the calculations of jet quenching, J/ψ melting, thermal photon and dilepton productions, etc. Thus hydro plays a central role in modeling relativistic heavy-ion collisions.

Hydrodynamics is formulated as an order-by-order expansion in gradients of u^μ, the ideal hydrodynamics being of the zeroth order. The zeroth-, first-, and second-order equations are named after Euler, Navier-Stokes, and Burnett, respectively, in the non-relativistic case (Fig. 1).

Journal of Physics: Conference Series **422** (2013) 012003 doi:10.1088/1742-6596/422/1/012003

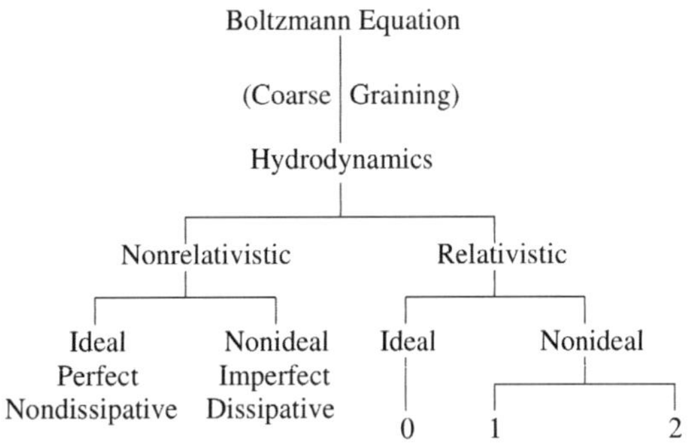

Figure 1. Coarse-graining of kinetic theory.

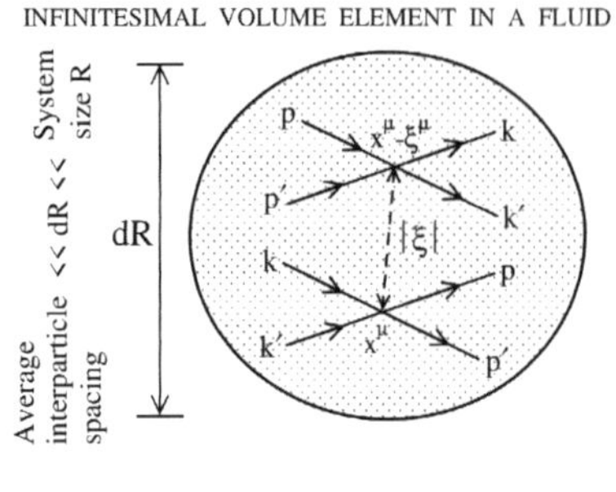

Figure 2. Collisions $kk' \to pp'$ and $pp' \to kk'$ occurring at points x^μ and $x^\mu - \xi^\mu$ within an infinitesimal fluid element of size dR, containing a large number of particles represented by dots.

The relativistic Navier-Stokes equations are parabolic in nature and exhibit acausal behaviour, which was rectified in the (second-order) Israel-Stewart (IS) theory [1]. The formulation of the relativistic dissipative second-order hydrodynamics ("2" in Fig. 1) is currently under intense investigation [2, 3, 4, 5, 6, 7, 8, 9, 10].

Hydrodynamics has traditionally been derived either from entropy considerations (i.e., the generalized second law of thermodynamics) or by taking the second moment of the Boltzmann equation; for a review, see [11]. The former approach captures a subset of the terms allowed by symmetry [12] in the evolution equations for the dissipative quantities. The latter approach captures some more terms but not all. The question 'why do the traditional approaches not generate *all* the allowed terms', was the main motivation of our work [13].

2. Present work

In [13] we presented a new derivation of the dissipative hydrodynamic equations within kinetic theory but using a nonlocal collision term in the Boltzmann equation. We obtained all the second-order terms that are allowed by symmetry [12] and showed that the coefficients of the existing terms in the widely used traditional IS equations were altered. These modifications do have a rather strong influence on the evolution of the viscous medium as we demonstrated in the case of one-dimensional scaling expansion.

It is important to recall at the outset that an infinitesimal volume element in the fluid is always supposed to be large compared with the mean interparticle spacing and hence contains "a very great" number of particles [14]; see also Fig. 2. In kinetic theory, the single-particle phase-space distribution function $f(x,p)$ is assumed to vary slowly over space-time, i.e., it changes negligibly over the range of interparticle interaction [15], which in fact is much smaller than the mean interparticle spacing. It is important to keep this hierarchy of length scales in mind.

It is also interesting to recall Israel and Stewart's classic paper [1] where they list the properties of the collision term $C[f]$. We quote: "We require only the following general properties: (i) $C[f]$ is a purely local function or functional of f, independent of $\partial_\mu f$. (ii) The form of $C[f]$ is consistent with conservation of 4-momentum and number of particles at collisions. (iii) $C[f]$ yields a non-negative expression for the entropy production .... These requirements are of course met by Boltzmann's ansatz for 2-particles collisions, and, indeed, one may hope that they hold somewhat more generally, although the locality assumption (i) is a powerful restriction." Note the words *hope* and *assumption*. The locality assumption is questionable and we relaxed it on the length scale of dR [13]. Despite the long history of the Boltzmann equation, a collision term containing $\partial_\mu f$, to our knowledge, has never been used to derive hydrodynamic equations. Such a collision term brings about a change not only in hydrodynamics but also in the kinetic theory.

Our starting point is the relativistic Boltzmann equation with the modified collision term:

$$p^\mu \partial_\mu f = C_m[f] = C[f] + \partial_\mu (A^\mu f) + \partial_\mu \partial_\nu (B^{\mu\nu} f) + \cdots, \tag{1}$$

where A^μ and $B^{\mu\nu}$ depend on the type of the collisions $(2 \leftrightarrow 2,\ 2 \leftrightarrow 3, \ldots)$.

For instance, for $2 \leftrightarrow 2$ elastic collisions,

$$C[f] = \frac{1}{2} \int dp' dk\ dk'\ W_{pp' \to kk'}\ (f_k f_{k'} \tilde{f}_p \tilde{f}_{p'} - f_p f_{p'} \tilde{f}_k \tilde{f}_{k'}), \tag{2}$$

where $W_{pp'\to kk'}$ is the transition rate, $f_p \equiv f(x,p)$ and $\tilde{f}_p \equiv 1 - rf(x,p)$ with $r = 1, -1, 0$ for Fermi, Bose, and Boltzmann gas, and $dp = g d\mathbf{p}/[(2\pi)^3 \sqrt{\mathbf{p}^2 + m^2}]$, g and m being the degeneracy factor and particle rest mass. The first and second terms in Eq. (2) refer to the gain and loss processes $kk' \to pp'$ and $pp' \to kk'$, respectively, occurring *anywhere* in the infinitesimal fluid element located at the space-time point x^μ. These processes have traditionally been assumed to occur at the same point x^μ with an underlying assumption that $f(x,p)$ is constant not only over the range of interparticle interaction but also over the entire fluid element of size dR. Boltzmann equation together with this crucial assumption has been used to derive the standard second-order dissipative hydrodynamic equations [1, 9, 11]. We, however, emphasize that the variation of $f(x,p)$ over the span of the fluid element may not be negligible, and hence the space-time points at which the above two kinds of processes occur should be separated by an interval $|\xi^\mu| \leq dR$ within the volume d^4R (Fig. 2). The large number of particles within d^4R collide among themselves with various separations ξ^μ. Of course, the points $(x^\mu - \xi^\mu)$ must lie within the past light-cone of the point x^μ (i.e., $\xi^2 > 0$ and $\xi^0 > 0$) to ensure that the evolution of $f(x,p)$ in Eq. (1) does not violate causality. With this realistic viewpoint, the second term in Eq. (2) involves $f(x-\xi,p) f(x-\xi,p') \tilde{f}(x-\xi,k) \tilde{f}(x-\xi,k')$, which on Taylor expansion at x^μ up to second order in ξ^μ, results in the modified Boltzmann equation (1) with

$$A^\mu = \frac{1}{2} \int dp' dk\ dk'\ \xi^\mu W_{pp'\to kk'} f_{p'} \tilde{f}_k \tilde{f}_{k'}, \ \text{and}\ B^{\mu\nu} = -\frac{1}{4} \int dp' dk\ dk'\ \xi^\mu \xi^\nu W_{pp'\to kk'} f_{p'} \tilde{f}_k \tilde{f}_{k'}. \tag{3}$$

In general, for all collision types $(2 \leftrightarrow 2,\ 2 \leftrightarrow 3, \ldots)$, the momentum dependence of the coefficients A^μ and $B^{\mu\nu}$ can be made explicit by expressing them in terms of the available tensors p^μ and the metric $g^{\mu\nu}$ as $A^\mu = a(x) p^\mu$ and $B^{\mu\nu} = b_1(x) g^{\mu\nu} + b_2(x) p^\mu p^\nu$. Equation (1) with this A^μ and $B^{\mu\nu}$ forms the basis of our derivation of the second-order dissipative hydrodynamics. Arguments in the previous paragraph were meant only to provide a physical motivation for the mathematical form of $C_m[f]$ in Eq. (1).

Conservation of current, $\partial_\mu N^\mu = 0$ and energy-momentum tensor, $\partial_\mu T^{\mu\nu} = 0$, yield the fundamental evolution equations for n, ϵ and u^μ (defined in the Landau frame)

$$\begin{aligned} Dn + n\partial_\mu u^\mu + \partial_\mu n^\mu &= 0, \\ D\epsilon + (\epsilon + P + \Pi)\partial_\mu u^\mu - \pi^{\mu\nu}\nabla_{(\mu}u_{\nu)} &= 0, \\ (\epsilon + P + \Pi)Du^\alpha - \nabla^\alpha(P+\Pi) + \Delta^\alpha_\nu \partial_\mu \pi^{\mu\nu} &= 0. \end{aligned} \tag{4}$$

Equations (4) together with the EoS constitute six equations in fifteen unknowns. How to derive the extra nine equations that would give us a closed set of equations? Boltzmann equation provides a way: The requirement of the conservation of energy-momentum and current implies vanishing zeroth and first moments of the collision term $C_m[f]$ in Eq. (1), i.e., $\int dp\ C_m[f] = 0 = \int dp\ p^\mu C_m[f]$ at each order in ξ^μ. In order to obtain the evolution equations for the dissipative quantities, we follow the IS approach [1] and consider the second moment of the modified Boltzmann equation (1)

$$\int dp\ p^\alpha p^\beta p^\gamma \partial_\gamma f = \int dp\ p^\alpha p^\beta [C[f] + p^\gamma \partial_\gamma (af) + \partial^2 (b_1 f_0) + (p \cdot \partial)^2 (b_2 f_0)], \tag{5}$$

and then take recourse to Grad's 14-moment approximation [16] for the single-particle distribution in orthogonal basis [9]. This gives the desired equations:

$$\begin{aligned} \Pi &= \tilde{a}\Pi_{\rm NS} - \beta_{\dot{\Pi}}\tau_\Pi \dot{\Pi} + \tau_{\Pi n} n \cdot \dot{u} - l_{\Pi n}\partial \cdot n - \delta_{\Pi\Pi}\Pi\theta + \lambda_{\Pi n} n \cdot \nabla\alpha + \lambda_{\Pi\pi}\pi_{\mu\nu}\sigma^{\mu\nu} \\ &\quad + \Lambda_{\Pi\dot{u}}\dot{u} \cdot \dot{u} + \Lambda_{\Pi\omega}\omega_{\mu\nu}\omega^{\nu\mu} + (8\ \text{terms}), \end{aligned} \tag{6}$$

$$\begin{aligned} n^\mu &= \tilde{a} n^\mu_{\rm NS} - \beta_{\dot{n}}\tau_n \dot{n}^{\langle\mu\rangle} + \lambda_{nn} n_\nu \omega^{\nu\mu} - \delta_{nn} n^\mu\theta + l_{n\Pi}\nabla^\mu\Pi - l_{n\pi}\Delta^{\mu\nu}\partial_\gamma \pi^\gamma_\nu - \tau_{n\Pi}\Pi\dot{u}^\mu \\ &\quad - \tau_{n\pi}\pi^{\mu\nu}\dot{u}_\nu + \lambda_{n\pi} n_\nu \pi^{\mu\nu} + \lambda_{n\Pi}\Pi n^\mu + \Lambda_{n\dot{u}}\omega^{\mu\nu}\dot{u}_\nu + \Lambda_{n\omega}\Delta^\mu_\nu \partial_\gamma \omega^{\gamma\nu} + (9\ \text{terms}), \end{aligned} \tag{7}$$

$$\begin{aligned} \pi^{\mu\nu} &= \tilde{a}\pi^{\mu\nu}_{\rm NS} - \beta_{\dot{\pi}}\tau_\pi \dot{\pi}^{\langle\mu\nu\rangle} + \tau_{\pi n} n^{\langle\mu}\dot{u}^{\nu\rangle} + l_{\pi n}\nabla^{\langle\mu} n^{\nu\rangle} + \lambda_{\pi\pi}\pi_\rho^{\langle\mu}\omega^{\nu\rangle\rho} - \lambda_{\pi n} n^{\langle\mu}\nabla^{\nu\rangle}\alpha - \tau_{\pi\pi}\pi_\rho^{\langle\mu}\sigma^{\nu\rangle\rho} \\ &\quad - \delta_{\pi\pi}\pi^{\mu\nu}\theta + \Lambda_{\pi\dot{u}}\dot{u}^{\langle\mu}\dot{u}^{\nu\rangle} + \Lambda_{\pi\omega}\omega_\rho^{\langle\mu}\omega^{\nu\rangle\rho} + \chi_1 \dot{b}_2 \pi^{\mu\nu} + \chi_2 \dot{u}^{\langle\mu}\nabla^{\nu\rangle} b_2 + \chi_3 \nabla^{\langle\mu}\nabla^{\nu\rangle} b_2, \end{aligned} \tag{8}$$

where $\tilde{a} = (1 - a)$ and $\dot{X} = DX$. The "8 terms" ("9 terms") involve second-order, linear scalar (vector) combinations of derivatives of b_1, b_2. All the terms in the above equations are inequivalent, i.e., none can be expressed as a combination of others via equations of motion [12]. All the coefficients in Eqs. (6)-(8) can be written as functions of hydrodynamic variables [13].

In [13], we demonstrated the numerical significance of the new dissipative equations derived here, by considering the evolution of a massless Boltzmann gas, with the equation of state $\epsilon = 3P$, at vanishing net baryon number density, in the Bjorken model [17].

3. Summary

To summarize, we have presented a new derivation of the relativistic dissipative hydrodynamic equations by introducing a nonlocal generalization of the collision term in the Boltzmann equation. The first-order (Navier-Stokes) and second-order (Israel-Stewart) equations are modified: new terms occur and coefficients of others are altered. While it is well known that the derivation based on the generalized second law of thermodynamics misses some terms in the second-order equations, we have shown that the standard derivation based on kinetic theory also misses other terms. The method presented here is able to generate all possible terms to a given order that are allowed by symmetry.

[1] Israel W and Stewart J M 1979 *Annals Phys.* **118** 341
[2] Baier R, Romatschke P and Wiedemann U A 2006 *Phys. Rev.* C **73** 064903
[3] Baier R, Romatschke P, Son D T, Starinets A O and Stephanov M A 2008 *J. High Energy Phys.* JHEP04(2008)100
[4] Bhattacharyya S, Hubeny V E, Minwalla S and Rangamani M 2008 *J. High Energy Phys.* JHEP02(2008)045
[5] Natsuume M and Okamura T 2008 *Phys. Rev.* D **77**, 066014 [Erratum 2008 *Phy. Rev.* D **78** 089902]
[6] Bhalerao R S and Gupta S 2008 *Phys. Rev.* C **77** 014902
[7] El A, Muronga A, Xu Z and Greiner C 2009 *Phys. Rev.* C **79** 044914
[8] El A, Xu Z and Greiner C 2010 *Phys. Rev.* C **81** 041901
[9] Denicol G S, Koide T and Rischke D H 2010 *Phys. Rev. Lett.* **105** 162501
[10] Denicol G S, Niemi H, Molnar E and Rischke D H 2012 *Phys. Rev.* D **85** 114047
[11] Romatschke P 2010 *Int. J. Mod. Phys.* E **19** 1
[12] Bhattacharyya S 2012 *J. High Energy Phys.* JHEP07(2012)104
[13] Jaiswal A, Bhalerao R S and Pal S 2012 *Preprint* 1204.3779 [nucl-th] (revised)
[14] Landau L D and Lifshitz E M 1987 *Fluid Mechanics* (Oxford: Butterworth-Heinemann) p 1
[15] de Groot S R, van Leeuwen W A and van Weert Ch G 1980 *Relativistic Kinetic Theory — Principles and Applications* (Amsterdam: North-Holland) p 18
[16] Grad H 1949 *Comm. Pure Appl. Math.* **2** 331
[17] Bjorken J D 1983 *Phys. Rev.* D **27** 140

Journal of Physics: Conference Series **422** (2013) 012004 doi:10.1088/1742-6596/422/1/012004

Higher Harmonics in Heavy Ion Collisions

Sangyong Jeon

Department of Physics, McGill University, 3600 University Street, Montreal, QC, Canada

E-mail: jeon@physics.mcgill.ca

Abstract. As the QGP expands and cools, it carries much information on its creation and evolution imprinted on the patterns of higher harmonic flow. In this proceeding we report on the progress in simulating and understanding the higher harmonics by the McGill group using the 3+1D event-by-event viscous hydrodynamics simulation suite named MUSIC.

1. Introduction

One of the major goals of heavy ion collision phenomenology is to study the creation and the subsequent evolution of the Quark-Gluon Plasma (QGP). Accomplishing this goal requires a remarkable range of theoretical tools, from classical and quantum field theory to Monte-Carlo simulations of hadronic collisions. Currently, one of the most important theoretical tools is the relativistic hydrodynamics.

While the whole of heavy ion collision may take around 15 fm/c, the system spends a bulk of that time in the hydrodynamic phase. Hydrodynamics studies assume that after a brief initial thermalization phase (on the order of 1 fm/c), hydrodynamics begins to apply. The properties of QGP such as the equation of state and the viscosities then critically influence subsequent evolution of the system. Hydrodynamics is essentially the response of a fluid to its local pressure gradient. Therefore, the imprint of the QGP properties needs to be sought in the fluid properties reflected in the final state hadrons. More explicitly, in the *global shape* of the momentum distribution of hadrons.

Mathematically, the shape is encoded in the coefficients of the Fourier series

$$\frac{dN}{dyd^2p_T} = \frac{1}{2\pi}\frac{dN}{dyp_Tdp_T}\left(1 + \sum_{n=1}^{\infty} 2v_n \cos(n(\phi - \psi_n))\right) \tag{1}$$

where ψ_n is the n-th order event-plane angle defined as

$$\psi_n = \frac{1}{n}\tan^{-1}\left(\frac{\langle p_T \sin(n\phi)\rangle}{\langle p_T \cos(n\phi)\rangle}\right). \tag{2}$$

It is customary to choose the orientation in such a way that $v_1 = 0$. The second coefficient v_2 is usually referred to as the elliptic flow and the third coefficient v_3, the triangular flow. The elliptic flow is a measure of how efficiently hydrodynamics translates the spatial anisotropy to the momentum anisotropy. Triangular flow depends more on the spatial distribution of the initial energy density fluctuations. Higher harmonic coefficients encode more detailed information on the initial fluctuations and their propagation through the QGP evolution. Therefore, how well one understands these flow coefficients is an indication of how well one understands the bulk properties of QGP.

Hydrodynamic studies of heavy ion collisions have a long history [1]. Until recently, most of the study was conducted using the boost-invariant 2+1D ideal hydrodynamics with smooth averaged initial conditions (For reviews, see Refs.[2, 3]). More recent efforts have generalized this to include viscosity [4, 5], 3+1D dynamics [6, 7], and fluctuating initial conditions [8]. However, none of the above mentioned studies has implemented all three improvements over the 2+1D ideal hydrodynamics. Only recently, the first implementation of 3+1D event-by-event viscous hydrodynamics was made by the McGill group [9, 10, 11]. In this proceeding, I report on recent progresses made by the McGill group based mostly on Ref.[11].

2. 3+1D Event-by-Event Viscous Hydrodynamics

Our implementation of hydrodynamics (named MUSIC) uses the hyperbolic τ-η_s coordinate system but without assuming the invariance in η_s. For the evolution of the viscous tensor $\pi^{\mu\nu}$, we use a variant of the Israel-Stewart formalism [12]

$$\Delta^{\mu\alpha}\Delta^{\nu\beta}u^\gamma\partial_\gamma\pi_{\alpha\beta} = -\frac{1}{\tau_\pi}\left(\pi^{\mu\nu} - 2\eta\nabla^{\langle\mu}u^{\nu\rangle} + \frac{4}{3}\tau_\pi\pi^{\mu\nu}(\partial_\alpha u^\alpha)\right) \tag{3}$$

where u^μ is the flow velocity and $\Delta^{\mu\nu} = g^{\mu\nu} - u^\mu u^\nu$ is the local 3-metric. For solving conservation equations, we use a recently developed algorithm of Kurganov and Tadmor [13] in the semi-discrete form together with the Heun's method. This is relatively simple to implement yet very stable. The fluctuating initial condition is implemented by using the Glauber approximation. Namely, the nucleons are first distributed in the projectile and the target nuclei according to the Wood-Saxon distribution function. Then according to their transverse position and the nucleon-nucleon cross-section the colliding pairs are chosen. At each collision point, a Gaussian shaped energy density is assigned. The width of the Gaussian is chosen not to exceed $c\tau_0$ where τ_0 is the initial time when hydrodynamic evolution begins. The energy density is adjusted to reproduce the final state hadron entropy.

For the equation of state, our choice is the lattice based parameterization "s95p-v1" from [14] although other choices are also available. Freeze-out space-time point for each cell at (x, y, η_s) is given by

$$\Sigma = (\tau_f(x, y, \eta_s)\cosh\eta_s, x, y, \tau_f(x, y, \eta_s)\sinh(\eta_s)) \tag{4}$$

where τ_f is the freeze-out time when the local temperature reaches the freeze-out temperature. Cooper-Frye formula then is used with the normal vector given by

$$d^3\Sigma_\mu = -\epsilon_{\mu\nu\lambda\rho}\frac{\partial\Sigma^\nu}{\partial x}\frac{\partial\Sigma^\nu}{\partial y}\frac{\partial\Sigma^\nu}{\partial\eta_s}dxdyd\eta_s \tag{5}$$

to get the final state hadronic spectrum with the full resonance decays adapted from Heinz and Kolb's 2-D scheme. The viscous correction to the thermal spectrum is given by

$$\delta f = f_0(1 \pm f_0)\frac{p^\alpha p^\beta\pi_{\alpha\beta}}{2sT^3} \tag{6}$$

where the $\pm$ signs are for Bosons (+) and Fermions (−) and s is the entropy density. Up to $p^\alpha = O(T)$, this is $O(\eta/s)$ or smaller.

3. Results and Discussion

In our previous publications, we have shown that our implementation of the hydrodynamics reproduces very well the measured elliptic flows at RHIC with the shear viscosity to entropy

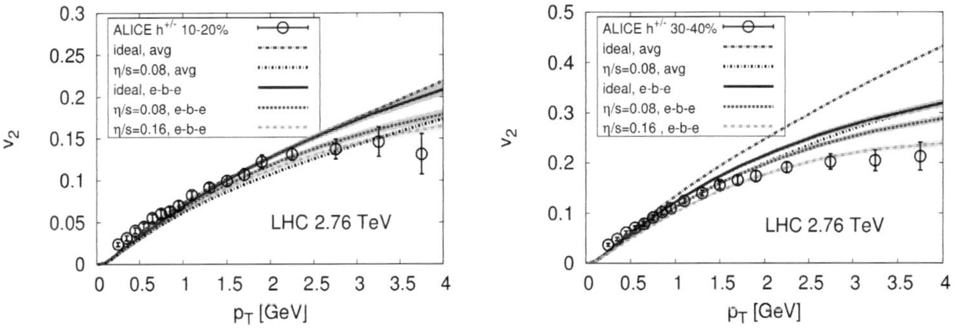

Figure 1. The differential elliptic flow at the LHC. The figures are taken from [11]. The Data points are from [15].

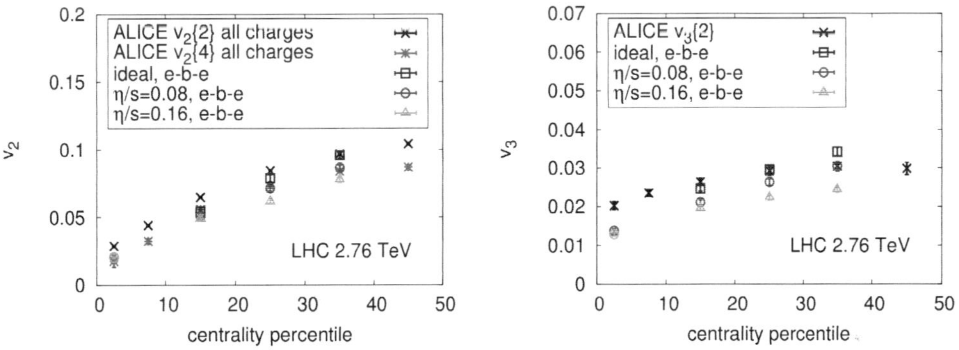

Figure 2. The integrated elliptic flow at LHC. The differential elliptic flow at the LHC. The figures are taken from [11]. The Data points are from [15].

ratio of $\eta/s = 1/4\pi$. We have also predicted a sizable triangular flow due to the initial state fluctuations which has been confirmed by the experiment [16] indicating that at RHIC, the ratio η/s is indeed very small. Our calculations for the LHC are shown in Figures 1 and 2. Shown in Figure 1 are the results for the elliptic flow for the 10–20 % centrality class and the 30–40 % centrality class. For more central collisions, it does not make much difference whether the average initial condition is used or the fluctuating initial condition is used. This is because for more central collisions, the nucleon-nucleon collision sites are close together. Therefore a single event does not look much different from the average one. One can also see that it makes little difference whether $\eta/s = 1/4\pi$ or $\eta/s = 1/2\pi$. This is due to the fact that the eccentricity is not so large in this centrality class. The short amount time it takes for the system to cease developing anisotropic flow is in effect too short for the difference in the viscosity to manifest.

For the 30–40 % centrality class, average initial condition and fluctuating initial condition lead to fairly different results. This is due to the fact that in this centrality class, the nucleon-nucleon collision sites are far enough apart that the average eccentricity tends to be smaller in the event-by-event cases. It is also evident from these figures that η/s at the LHC is bigger than that at RHIC by a factor of 2. This is not unexpected since higher temperature at the LHC would imply smaller α_S which in turn would imply higher η/s.

Shown in Figure 2 are our calculations of the integrated v_2 and v_3 compared to the ALICE

Journal of Physics: Conference Series **422** (2013) 012004 doi:10.1088/1742-6596/422/1/012004

data. We underestimate v_2 and v_3 slightly. This may be due to the fact that the Glauber initial condition tends to underestimate the eccentricity. More study is underway.

4. Conclusions and Outlook

Using the first implementation of 3+1D event-by-event viscous hydrodynamics, we have begun to investigate the higher harmonic components of the QGP flow. As the experiments accumulate more data, the higher harmonics can potentially provide a very discriminating probe of the nature of QGP and its evolution throughout relativistic heavy ion collisions. Our investigation with the Glauber initial conditions shows that the viscosity parameter at the LHC is about twice as big as the viscosity parameter at RHIC in accordance with the asymptotic freedom property of QCD. Studies with other initial conditions such as the IP-Sat model are underway. Efforts to make improvements such as the hadronic afterburner and jet-fluid interactions are also underway.

References

[1] Belenkij S and Landau L 1956 *Nuovo Cim.Suppl.* **3S10** 15
[2] Kolb P F and Heinz U W 2004 *Quark Gluon Plasma 3* ed Hwa R C and Wang X N (Singapore: World Scientific) (*Preprint* nucl-th/0305084)
[3] Huovinen P 2004 *Quark Gluon Plasma 3* ed Hwa R C and Wang X N (Singapore: World Scientific) (*Preprint* nucl-th/0305064)
[4] Heinz U W, Song H and Chaudhuri A K 2006 *Phys.Rev.* **C73** 034904 (*Preprint* nucl-th/0510014)
[5] Baier R and Romatschke P 2007 *Eur.Phys.J.* **C51** 677–687 (*Preprint* nucl-th/0610108)
[6] Muroya S and Nonaka C 1997 (*Preprint* nucl-th/9709004)
[7] Hirano T 2001 *Phys.Rev.Lett.* **86** 2754–2757 (*Preprint* nucl-th/0004029)
[8] Aguiar C, Hama Y, Kodama T and Osada T 2002 *Nucl.Phys.* **A698** 639–642 (*Preprint* hep-ph/0106266)
[9] Schenke B, Jeon S and Gale C 2010 *Phys.Rev.* **C82** 014903 (*Preprint* 1004.1408)
[10] Schenke B, Jeon S and Gale C 2011 *Phys.Rev.Lett.* **106** 042301 (*Preprint* 1009.3244)
[11] Schenke B, Jeon S and Gale C 2011 *Phys.Lett.* **B702** 59–63 (*Preprint* 1102.0575)
[12] Baier R, Romatschke P, Son D T, Starinets A O and Stephanov M A 2008 *JHEP* **0804** 100 (*Preprint* 0712.2451)
[13] Kurganov A and Tadmor E 2000 *Journal of Computational Physics* **160** 241
[14] Huovinen P and Petreczky P 2010 *Nucl.Phys.* **A837** 26–53 (*Preprint* 0912.2541)
[15] Aamodt K *et al.* (ALICE Collaboration) 2010 *Phys.Rev.Lett.* **105** 252302 (*Preprint* 1011.3914)
[16] Adare A *et al.* (PHENIX Collaboration) 2011 *Phys.Rev.Lett.* **107** 252301 (*Preprint* 1105.3928)

Journal of Physics: Conference Series **422** (2013) 012005 doi:10.1088/1742-6596/422/1/012005

Global and Collective Dynamics at PHENIX

Takafumi Niida for the PHENIX Collaboration
University of Tsukuba, 1-1-1 Tennoudai, Tsukuba, Ibaraki 305-8571, Japan
E-mail: niida@rcf.rhic.bnl.gov

Abstract. In order to study the properties of the hot and dense matter created by heavy ion collisions, various physics observables have been measured in RHIC and the LHC, such as spectra, collective flow, two particle correlations and HBT. The higher harmonic flow v_n have been recently measured, which is primarily coming from the spatial fluctuation of the initial participant density. Measurements related to the v_n and event planes will give us new insight into the properties of the hot and dense matter and space-time evolution in the heavy ion collisions. We present the latest results of higher harmonic flow v_n for identified particles and azimuthal hadron correlations with the subtraction of backgrounds from higher harmonic flow v_n. Besides, we present m_T dependence of final eccentricity obtained by azimuthal HBT measurement.

1. Introduction

The initial geometry of the overlap region in heavy ion collisions have been treated as the collisions of nuclei with smooth density profile so far. In recent study, it is considered that the spatial position of the participating nucleons fluctuates geometrically and it leads to higher harmonic deformation. This spatial fluctuation is converted to momentum space by the collective expansion and then higher harmonic flow v_n is created.

Azimuthal distribution of emitted particles is written as the following:

$$\frac{dN}{d\phi} \propto 1 + \sum_{n=1} 2v_n \cos(\mathrm{n}(\phi - \Phi_\mathrm{n})) \quad (1)$$

$$v_n = \langle \cos(\mathrm{n}(\phi - \Phi_\mathrm{n})) \rangle \quad (2)$$

where n indicates the harmonic order, ϕ and Φ_n denote the azimuthal angle of emitted particles and event plane for each harmonic orders. The flow strength v_n is represented as the Fourier coefficient. The event planes are determined by the Reaction Plane Detector which covers $1.0 < |\eta| < 2.8$ and charged particles are measured by the PHENIX central detectors with $|\eta| < 0.35$ in this analysis. Charged hadron v_n have been measured at PHENIX [1], where v_2 increases from central to peripheral collisions, while v_3 doesn't seems to have centrality dependence. It indicates that v_3 mainly comes from the initial fluctuation. Though some theoretical models with different initial conditions can explain v_2 well, some of them fail to explain v_3. So v_3 provides the new constraints on the theoretical models. It is expected that v_n for identified particles will provide more constraints on the theoretical models.

2. Higher harmonic flow

Figure 1 shows higher harmonic flow of $\pi^{\pm}$, $K^{\pm}$ and $p(\bar{p})$ as a function of p_T in 0-50% in $\sqrt{s_{\mathrm{NN}}}$ = 200 GeV Au+Au collisions. The measured $v_4(\Phi_2)$ with respect to 2nd order event plane is also shown. Green bands indicate p_T correlated systematic uncertainties. The shown $v_3(\Phi_3)$, $v_4(\Phi_4)$ and $v_4(\Phi_2)$ are multiplied by factor, 1.5, 1.5 and 5.0 respectively when plotted in Fig.1. Each v_n for different species shows same trends. Particle mass dependence at low p_T and baryon/meson difference at mid p_T can be seen in v_3, v_4 as well as v_2. It is considered that higher harmonic flow shows the behavior like the hydrodynamics at low p_T region, and described by the quark recombination model at mid p_T region as well as v_2.

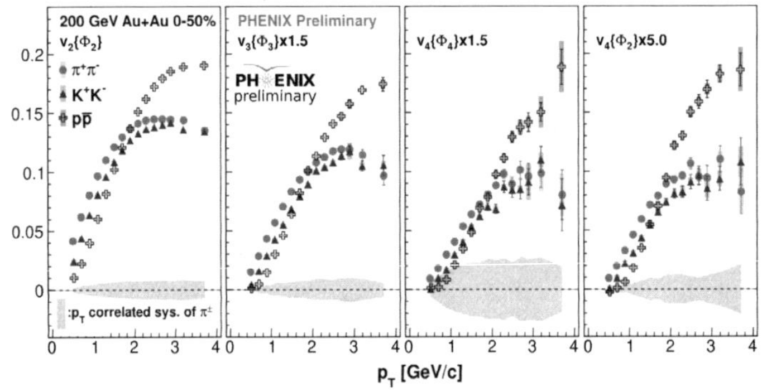

Figure 1. (Color online)$v_2(\Phi_2), v_3(\Phi_3), v_4(\Phi_4)$ and $v_4(\Phi_2)$ of $\pi^{\pm}, K^{\pm}$ and $p(\bar{p})$ as a function of p_T in 0-50%. Green bands indicate p_T correlated systematic uncertainties. v_3 and v_4 are plotted after being multiplied by scale factors.

Figure 2 shows the modified n_q scaling of higher harmonic flow as a function of KE_T/n_q, where n_q is the number of constituent quark, $KE_T = (p_T^2 + m_0^2)^{1/2} - m_0$, m_0 is the particle mass, and v_n is scaled by $n_q^{n/2}$ using higher harmonic order n. The modified n_q scaling succeeds to scale v_n. It indicates that higher harmonic flow v_n also arises at partonic level.

In the PHENIX experiment, charged pions and protons can be identified up to $p_T = 6$ GeV/c and kaons up to $p_T = 4$ GeV/c by combining Time-Of-Flight detector and Aerogel Cherenkov Counter. Figure 3 shows PID v_2 as functions of p_T, KE_T/n_q and p_T/n_q for two different centrality bins. KE_T/n_q scaling works better than p_T/n_q scaling in central collisions. However, it breaks for protons at $KE_T/n_q \approx 1.0$ GeV in non-central collisions. It is considered that particle production at $p_T > 4$ GeV/c is not dominated by quark recombination, but by jet energy loss. So this result may give us any information on the interplay between the recombination and jet energy loss at high p_T region.

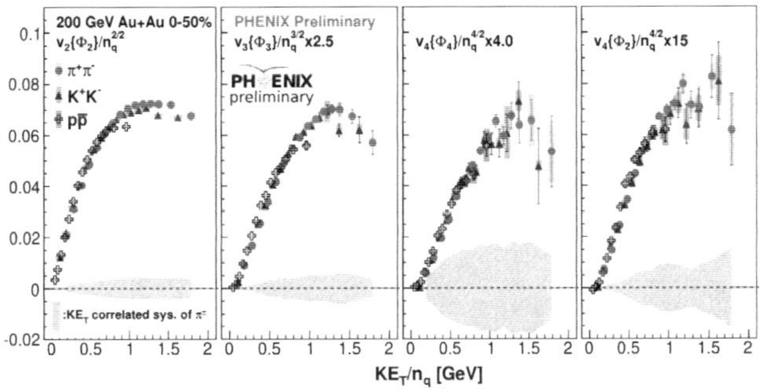

Figure 2. (Color online)$v_2(\Phi_2), v_3(\Phi_3), v_4(\Phi_4)$ and $v_4(\Phi_2)$ divided by $n_q^{n/2}$ of $\pi^{\pm}, K^{\pm}$ and $p(\bar{p})$ as a function of KE_T/n_q in 0-50%, where n and n_q denote harmonic order and the number of constituent quarks of hadrons. Green bands indicate p_T correlated systematic uncertainties.

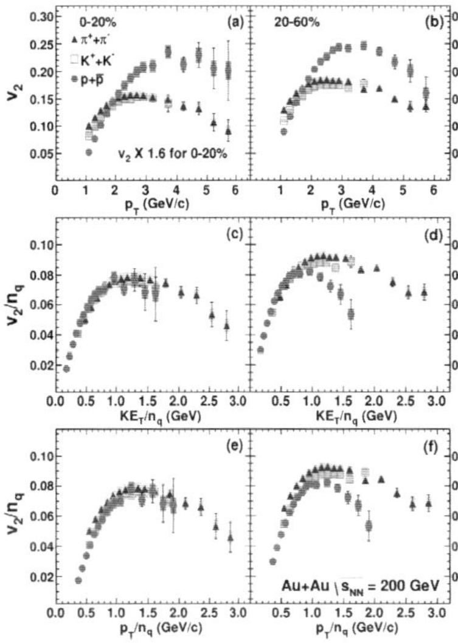

Figure 3. (Color online)v_2 of $\pi^{\pm}, K^{\pm}$ and $p(\bar{p})$ with and without n_q scaling

3. Two particle correlations

Two particle azimuthal correlations have been measured in heavy ion collisions to study the interaction between hard-scattered partons and the hot and dense matter. In recent studies, it is considered that higher harmonic flow will be a possible source of "ridge" and "mach cone" [2], which are the long range $\Delta\eta$ correlations at near side and double hump structure at away side. Figure 4 shows the recent result of the two particle correlations for charged hadrons with p_T^{trig}=2-4 GeV/c and p_T^{assoc}=1-2 GeV/c for $|\Delta\eta| < 0.7$ in $\sqrt{s_{\rm NN}}$ = 200 GeV Au+Au collisions, where v_2, v_3, v_4 backgrounds are subtracted by ZYAM method [3]. In most central collisions, away side yield is suppressed, while the yield and double hump structure still remain in moving to mid central collisions. It is difficult to explain the structure only by higher harmonic flow v_n.

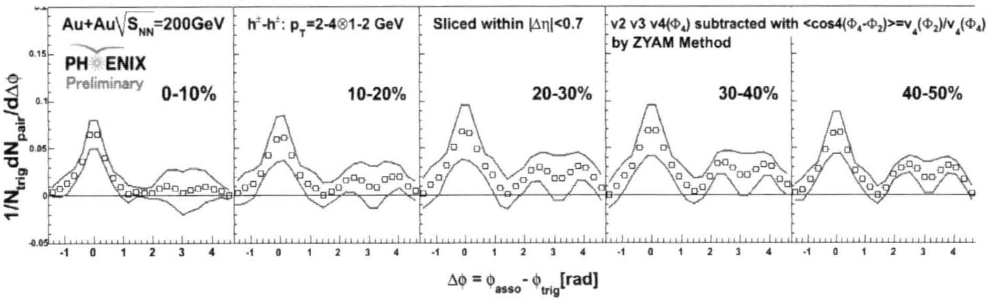

Figure 4. Two particle correlations for charged hadrons with p_T^{trig} =2-4 GeV/c and p_T^{assoc}=1-2 GeV/c for $|\Delta\eta| < 0.7$ in Au+Au 200 GeV collisions, where $v_2, v_3, v_4(\Phi_4)$ background are subtracted by ZYAM method [3].

4. Azimuthal HBT measurement

HBT measurement is a powerful tool to study the space-time evolution of particle emitting source. Azimuthal HBT radii with respect to 2nd-order event plane have been measured for charged pions and kaons to investigate the source shape at freeze-out at PHENIX [4]. The result shows that final eccentricity of kaons defined as $\varepsilon_{final} = 2R^2_{s,2}/R^2_{s,0}$ [5] is larger than that of pions and almost the same as the initial eccentricity. However, since HBT radii show a transverse mass (m_T) dependence and average m_T of pions and kaons are different, the m_T dependence needs to be considered in the comparison of the final eccentricities. Figure 5 shows the relative amplitude of azimuthal HBT radii of charged pions and kaons as a function of $\langle m_T \rangle$ for two centrality bins in $\sqrt{s_{\rm NN}} = 200$ GeV Au+Au collisions. Left top panel shows final eccentricity, and there is still difference between pions and kaons in non-central collisions even at the same $\langle m_T \rangle$. The difference may indicate faster freeze-out of kaons due to the lower cross section. The relative amplitude of R_o and R_{os} at low m_T in most central collisions have finite value though final eccentricity is close to zero. This result may indicate a temporal variation of emission duration of particles because R_o and R_{os} have temporal information in addition to geometrical information.

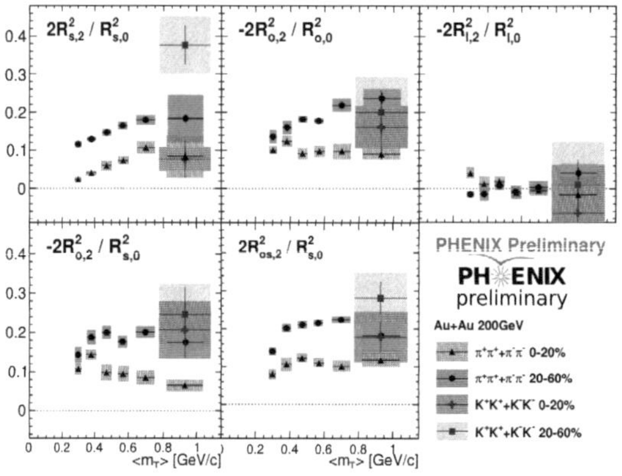

Figure 5. (Color online)Relative amplitude of azimuthal HBT radii of charged pions and kaons as a function of average m_T for 2 centrality bins. Left top panel corresponds to final eccentricity.

5. Summary

Recent results on higher harmonic flow for identified particles, two particle correlations with the background subtraction from higher harmonic flow and m_T dependence of azimuthal HBT measurement are presented. There are still some open questions about each of the topics discussed and further study in experiment and theory will be needed to understand them.

References

[1] A. Adare, *et al.* (PHENIX Collaboration), *Phys. Rev. Lett.* **107**, 252301 (2011)
[2] G. Aad *et al.* (ATLAS Collaboration), *Phys. Rev.* **C86**, 014907 (2012)
[3] S. S. Adler *et al.* (PHENIX Collaboration), *Phys. Rev. Lett.* **97**, 052301 (2006)
[4] T. Niida for the PHENIX Collaboration, in proceedings of *The Seventh Workshop on Particle Correlations and Femtoscopy* [PoS(WPCF2011)005]
[5] Fabrice Retière and Michael Annan Lisa, *Phys. Rev.* **C70**, 044907 (2004)

Journal of Physics: Conference Series **422** (2013) 012006 doi:10.1088/1742-6596/422/1/012006

Quarkonia production in relativistic heavy ion collisions[1]

Che Ming Ko, Kyongchol Han and Taesoo Song

Cyclotron Institute and Department of Physics and Astronomy, Texas A&M University, College Station, Texas 77843, USA

E-mail: ko@comp.tamu.edu; khan@comp.tamu.edu; songtsoo@yonsei.ac.kr

Abstract. Using the two-component model that includes both initial production from nucleon-nucleon hard scattering and regeneration from produced quark-gluon plasma (QGP), we have studied the effect of medium modifications of the binding energies and radii of quarkonia on their production in heavy-ion collisions. We find that the inclusion of medium effects is helpful for understanding the observed suppression of quarkonia production in experiments carried out at SPS, RHIC, and LHC.

Since the suggestion by Matsui and Satz [1] that the production of J/ψ would be suppressed in relativistic heavy-ion collisions as a result of the screening of color charges in QGP and thus could be used as a signature for the produced QGP in these collisions, there have been extensive experimental and theoretical studies on this very interesting phenomenon. Although experiments from SPS, RHIC, and LHC have indeed shown a suppressed production of J/ψ as well as the Υ, a satisfactory explanation involves the interplay of many effects from both the initial cold nuclear matter and the final hot partonic and hadronic matters. Including these effects in a two-component model that takes into account both initial production from nucleon-nucleon hard scattering and regeneration from produced QGP [2], we have recently carried out a series of studies of quarkonia production in relativistic heavy ion collisions [3, 4, 5].

In our studies, the numbers of initially produced quarkonia are obtained from the binary collisions among nucleons in the two colliding nuclei by using the empirical production cross sections and after including the shadowing effect via the EPS09 package [6] by assuming that it is proportional to the path length of a parton in a nucleus. The absorption effect in the cold nuclear matter is included for J/ψ by using a J/ψ-nucleon absorption cross section of 4.18 mb [7] and 2.8 mb [8] for SPS and RHIC, respectively, and 0 or 2.8 mb for LHC, but is neglected for Υ because of its large binding and small size. The local temperature of initially produced matter, which is needed for determining if quarkonia from nucleon-nucleon hard scattering can be formed, is estimated from the local entropy density $ds/d\eta = C[(1-\alpha)n_{\rm part}/2 + \alpha n_{\rm coll}]$ in terms of the number densities $n_{\rm part}$ and $n_{\rm coll}$ of participants and binary collisions, respectively. The parameters C and α are determined to be 14.6 and 0 for Pb+Pb collisions at 158A GeV at SPS, 18.7 and 0.11 for Au+Au collisions at $\sqrt{s_{NN}} = 200$ GeV at RHIC, and 27.0 and 0.15 for Pb+Pb collisions at $\sqrt{s_{NN}} = 2.76$ TeV at LHC from fitting measured multiplicity of final

[1] Based on work supported in part by the U.S. National Science Foundation under Grant No. PHY-1068572, the US Department of Energy under Contract No. DE-FG02-10ER41682, and the Welch Foundation under Grant No. A-1358.

Journal of Physics: Conference Series **422** (2013) 012006 doi:10.1088/1742-6596/422/1/012006

charged particles after hydrodynamic evolution. The latter is based on a schematic viscous hydrodynamics [9] using an equation of state given by the quasiparticle model of three flavors for the QGP and the resonance gas model for the hadron gas (HG) [10], together with the specific shear viscosity η/s taken to be 0.16 for the QGP at SPS and RHIC and 0.2 for that at LHC, and 0.8 for the hadron gas [11]. The hydrodynamic equations are solved using an initial thermalization time $\tau_0 = 1.0$, 0.9, and 1.05 fm/c for central collisions at SPS, RHIC, and LHC [12], respectively, with corresponding initial maximum temperatures of 224, 324 and 390 MeV and mean temperatures of 218, 269 and 311 MeV. We note that the longer initial thermalization time at LHC is to compensate for the larger viscosity, which would otherwise lead to too large a radial flow [12].

The in-medium properties of quarkonia in the QGP are determined by using the screened Cornell potential [13] between heavy quark and antiquark, i.e., $V(r,T) = (\sigma/\mu(T))[1 - e^{-\mu(T)r}] - (\alpha_s/r)e^{-\mu(T)r}$, with $\sigma = 0.192$ GeV^2, $\alpha_s = 0.471$, and the screening mass $\mu(T) = \sqrt{N_c/3 + N_f/6}\ gT$ in terms of the color number N_c, the light quark flavor number N_f, and the QCD coupling constant g. Compared to results from the lattice QCD calculations for a heavy quark and antiquark pair in the QGP [14], this potential is closer to their internal energy at high temperature but to their free energy at low temperature, thus interpolating smoothly the expected temperature-dependent potential between heavy quark and antiquark [15]. Taking the heavy quark masses m_c =1.32 GeV and m_b = 4.746 GeV [13] and g = 1.87, we solve the resulting Schrödinger equation for the heavy quark and antiquark pair to obtain the binding energies and radii of quarkonia in the QGP. The resulting dissociation temperatures are 1.76 T_c for J/ψ, 1.0 T_c for ψ' and χ_c, and 4.0, 1.67, 1.12, 1.51 and 1.09 T_c for Υ(1S), Υ(2S), Υ(3S), χ_b(1P) and χ_b(2P), respectively, where the critical temperature is taken to be T_c = 170 MeV.

For quarkonia that have survived in the QGP from initial dissociations, their number N_i of type i evolves in time according to the rate equation $dN_i/d\tau = -\Gamma_i(N_i - N_i^{\rm eq})$, where τ is the longitudinal proper time and Γ_i is the thermal decay width of quarkonia. The number of equilibrated quarkonia of type i is given by $N_i^{\rm eq} = \gamma^2 R\, n_i\, fV\theta(T_i - T)$, where n_i is their number density in grandcanonical ensemble; f is the fraction of QGP in the mixed phase; and $\theta(T_i - T)$ is the step function with T_i being the dissociation temperature of quarkonia of type i. The effect of chemical and kinetic off-equilibrium of heavy quarks in the QGP is included through their fugacity γ and relaxation factor R, with the former obtained from the conservation of heavy quark flavor [4] and the latter defined as $R(\tau) = 1 - \exp[-(\tau - \tau_0)/\tau_{\rm eq}]$ where $\tau_{\rm eq}$ is the relaxation time of heavy quarks [5]. With $\tau_{\rm eq} \sim m_Q$ [16], the relaxation time is about 4 fm/c for charm quarks [4, 17] and about 14 fm/c for bottom quarks. The thermal decay widths of quarkonia in QGP are calculated using the quarkonium dissociation cross sections $\sigma_i^{\rm diss}$ evaluated up to the next-to-leading order (NLO) in perturbative QCD [18], including thus dissociation by absorbing not only a thermal gluon but also the gluon emitted from a thermal quark or gluon. For the thermal width in the mixed phase, it is taken to be a linear combination of those in the QGP and the HG with the latter evaluated by using the quarkonia dissociation cross sections that are obtained from the factorization formula for the ground-state [18] and the assumption that those of excited stated are proportional to their squared radii. The thermal decay widths of quarkonia obtained with their in-medium binding energies and radii are found to increase with temperature and are significantly larger than those in free space.

In the left and middle windows of Fig. 1, we show by solid lines the calculated nuclear modification factor R_{AA} of J/ψ as a function of the participant number in Pb+Pb collisions at $\sqrt{s_{NN}}$ = 17.3 GeV at SPS, Au+Au collisions at $\sqrt{s_{NN}}$ = 200 GeV at RHIC, and Pb+Pb collisions at $\sqrt{s_{NN}}$ = 2.76 TeV at LHC (for transverse momentum larger than 6.5 GeV/c). They agree very well with the experimental data [7, 19, 20], which show a decreasing R_{AA} with increasing number of participants. Also shown are results from the primordial (dashed lines) and the regenerated J/ψ in the QGP (dotted lines), and they are seen to decrease and increase,

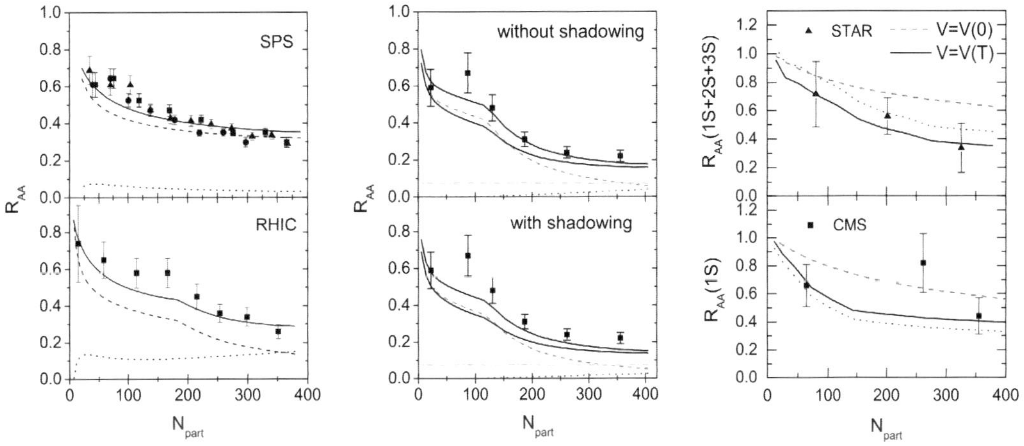

Figure 1. (Color online) Left window: Nuclear modification factor R_{AA} of J/ψ (solid lines) as a function of the participant number $N_{\rm part}$ in Pb+Pb collisions at $\sqrt{s_{NN}}$ = 17.3 GeV at SPS (upper panel) and Au+Au collisions at $\sqrt{s_{NN}}$ = 200 GeV at RHIC (lower panel). Middle window: Same for Pb+Pb collisions at $\sqrt{s_{NN}}$ = 2.76 TeV at LHC without (upper panel) and with (lower panel) the shadowing effect. Dashed and dotted lines represent, respectively, the contributions from primordial hard nucleon-nucleon scattering and regeneration in the QGP. Upper solid lines and dot-dashed lines in the right window are the R_{AA} of J/ψ at LHC for the nuclear absorption cross sections of $\sigma_{\rm abs}$ =0 and the contribution from decay of bottom hadrons, respectively. Right window: Same for the sum of Υ(1S), Υ(2S) and Υ(3S) at RHIC (upper panel) and that of Υ(1S) at LHC (lower panel). Solid and dashed lines are, respectively, results with and without medium effects on bottomonia. Dotted lines are results including also the shadowing or anti-shadowing effect. Experimental data are from Refs. [7, 19, 20, 23].

respectively, with increasing collision energy. For LHC, we have shown the results for both with (lower panel of middle window) and without (upper panel of middle window) the shadowing effect. The shadowing effect suppresses the production of charm pairs and consequently the regeneration of J/ψ. We have also included the contribution to J/ψ production from the decay of bottom hadrons at LHC [21, 22] by assuming that the R_{AA} of J/ψ from the decay of bottom hadrons is independent of the centrality [20]. As shown by the dash-dotted line in the middle window of Fig. 1, this contribution is comparable to that from the regenerated J/ψ (dotted lines) in peripheral collisions and more important than the primordial ones (dashed lines) in more central collisions. The lower solid lines in the middle window for LHC are the final R_{AA} of J/ψ obtained with a non-zero nuclear absorption cross section of 2.8 mb, and it differs from that without the nuclear absorption mainly in collisions of small number of participants as the primordial J/ψ are mostly dissolved in central and semi-central collisions.

In the upper panel of the right window of Fig. 1, the calculated $R_{\rm AA}$ of the sum of Υ(1S), Υ(2S) and Υ(3S) in Au+Au collisions at $\sqrt{s_{NN}}$ = 200 GeV at RHIC as a function of the participant number is shown and compared with the experimental data from the STAR Collaboration [23]. The solid and dotted lines are results obtained without and with the anti-shadowing effect in cold nuclear matter, and the latter is larger than the former. Because of the large experimental errors, both can describe the data from RHIC for all centralities. For Pb+Pb collisions at $\sqrt{s_{NN}}$ = 2.76 TeV at LHC, only the $R_{\rm AA}$ of $\Upsilon(1S)$ has been measured by the CMS Collaboration [20]. Our results without (solid line) and with (dotted line) the cold nuclear matter effect, shown in the lower panel, both agree with the data for peripheral (20-100%)

and most central (0-10%) collisions. For midcentral (10-20%) collisions, our model significantly underestimates the measured R_{AA}. We note that the suppression of $\Upsilon(1S)$ is largely due to the reduced feed-down contribution from its suppressed excited states. Also, the contribution from regeneration to the R_{AA} of bottomonia is less than 1% at both RHIC and LHC and for all centralities as a result of the small number of bottom quarks and the much longer bottom quark relaxation time than the lifetime of produced QGP. For comparison, results for the $R_{\rm AA}$ of bottomonia without medium effects are shown by dashed lines in the right window of Fig. 1, and they are larger than those with medium effects as expected and is thus large compared to the experimental data from RHIC, particularly for more central collisions. This is also the case for heavy ion collisions at the LHC except for midcentral collisions where the result obtained without medium effects can better describe the experimental data.

We have further studied the effects of initial state fluctuations on the R_{AA} of bottomonia at LHC by using the 2+1 dimension ideal hydrodynamics [5]. We find that while initial fluctuations hardly affect the yield of the 1S ground state bottomonium, their effect on that of excited bottomonium states is not small. Compared to the case with smooth initial conditions, the survival probabilities of excited bottomonium states are reduced at low transverse momentum in the case of large initial fluctuations while they are increased in the case of small initial fluctuations as a result of the smearing effect introduced in solving the hydrodynamic equations. The observed suppression of the excited bottomonia relative to the ground state bottomonium of an average transverse momentum of 6.57 GeV/c by CMS Collaborations can, however, be described at present with both smooth and fluctuating initial conditions.

In summary, using the two-component model that includes both initial production from nucleon-nucleon hard scattering and regeneration from produced quark-gluon plasma, we have studied quarkonia production in heavy-ion collisions at SPS, RHIC, and LHC by including the medium effects on the thermal properties of quarkonia and their dissociation cross sections. Our study shows that the inclusion of medium effects on quarkonia is helpful for describing the experimental observations.

References

[1] Matsui T and and Satz H 1986 *Phys. Lett.* B **178** 416
[2] Grandchamp L and Rapp R 2002 *Nucl. Phys.* **A709** 415
[3] Song T, Ko C M, Lee S H and Xu J 2011 *Phys. Rev.* C **83** 014914
[4] Song T, Han K C and Ko C M 2011 *Phys. Rev.* C **84** 034907; *ibid.* **85** 014902; *ibid* **85** 054905
[5] Song T, Han K C and Ko C M [arXiv:1112.0613 [nucl-th]]
[6] Eskola K J, Paukkunen H, Salgado C A 2009 *JHEP* **0904** 065
[7] Alessandro B *et al.* [NA50 Collaboration] 2005 *Eur. Phys. J.* C **39** 335
[8] Adare A *et al.* [PHENIX Collaboration] 2008 *Phys. Rev.* C **77** 024912 [Erratum-ibid. C **79**, 059901 (2009)]
[9] Song T, Han K C and Ko C M 2011 *Phys. Rev.* C **83** 024904
[10] Levai P and Heinz U W 1998 *Phys. Rev.* C **57** (1998).
[11] Demir N and Bass S A 2009 *Phys. Rev. Lett.* **102** 172302
[12] Song H, Bass S A and Heinz U 2011 *Phys. Rev.* C **83** 054912
[13] Karsch F, Mehr M T and Satz H 1988 *Z. Phys.* C **37** 617
[14] Kaczmarek O 2009 *Eur. Phys. J.* C **61** 811
[15] Wong C Y 2005 *Phys. Rev.* C **72** 034906
[16] Moore G D and Teaney D 2005 *Phys. Rev.* C **71** 064904
[17] Zhao X and Rapp R 2008 *Phys. Lett.* B **664** 253
[18] Song T and Lee S H 2005 *Phys. Rev.* D **72** 034002
[19] Adare A *et al.* [PHENIX Collaboration] 2007 *Phys. Rev. Lett.* **98** 232301
[20] CMS Collaboration 2011 *Quarkonium production in PbPb collisions at* $\sqrt{s_{NN}} = 2.76$ *TeV* CMS-PAS-HIN-10-006
[21] The ATLAS Collaboration [arXiv:1104.3038 [hep-ex]]
[22] Aaij R *et al.* [LHCb Collaboration] 2011 *Eur. Phys. J.* **C71** 1645
[23] Reed R [STAR Collaboration] [arXiv:1109.3891 [nucl-ex]]

Journal of Physics: Conference Series **422** (2013) 012007 doi:10.1088/1742-6596/422/1/012007

Collision Rate and Symmetry Factor in Gluon Plasma

Jian Deng
School of Physics, Shandong University, Shandong 250100, People's Republic of China

Qun Wang
Interdisciplinary Center for Theoretical Study and Department of Modern Physics, University of Science and Technology of China, Anhui 230026, People's Republic of China

Abstract. The elastic and inelastic collision rates in a gluon gas are calculated. The symmetry factor and the phase space integral are discussed in detail. With a symmetry factor and well constrained phase space, the same result can be obtained as that of the full phase space without the factor. Such an equivalence is illustrated by analytic and numerical calculations for $gg \to gg$ and $gg \to ggg$ processes.

It has been shown in AdS/CFT that the shear viscosity to entropy density ratio η/s in a physical system has a universal minimal bound $1/(4\pi)$ [1]. A recent perturbative QCD calculation [2, 3] for a gluon plasma shows that such a limit $\eta/s \sim 1/4\pi$ can be reached by the inelastic process $gg \leftrightarrow ggg$ (23), and the 23 process is 7 times more important than the $gg \leftrightarrow gg$ (22) process. It is in contradiction with the result by Arnold, Moore and Yaffe (AMY) [4, 5], where the 23 process only provides about a 10% correction to the 22 process. In our previous works [6, 7, 8], we have pointed out that such a discrepancy might arise from a multiple counting when carrying out collisional integrals of the 23 process over full phase space with a symmetry factor. In this note, we will provide a cross check about the above implication by calculating collision rates with and without symmetry factors. We will illustrate the equivalence of two ways of carrying out collisional integrals, in the constrained phase space with the symmetry factor or in full phase space without the symmetry factor.

For a hot gluon plasma in the vicinity of thermal equilibrium, the collision rates are defined by the detailed balance rate in Boltzmann equation [6]. For the 22 process, due to the symmetry between the u-channel and t-channel in the matrix element $|M_{12\to 34}|^2$, we may choose one channel in the the phase space integration and multiply the result by a symmetry factor of 2. For example, if we choose the t-channel and limit the phase space to the near forward angle scatterings, where $\mathbf{q} = \mathbf{p}_3 - \mathbf{p}_1$, and $\mathbf{q}_T$ is the transverse component of $\mathbf{q}$ with respect to $\mathbf{p}_1$ in the CM frame, the matrix element reads

$$|M_{12\to 34}|^2_{CM} \underset{\mathbf{q}^2\approx \mathbf{q}_T^2\approx 0}{\approx} 144\pi^2\alpha_s^2 \frac{s^2}{(\mathbf{q}_T^2 + m_D^2)^2}. \quad (1)$$

Here $\alpha_s = g^2/(4\pi)$ is the strong coupling constant, and we have inserted the Debye mass $m_D \sim gT$ to regularize gluon propagator. If we do not limit the phase space in the integration,

Journal of Physics: Conference Series **422** (2013) 012007 doi:10.1088/1742-6596/422/1/012007

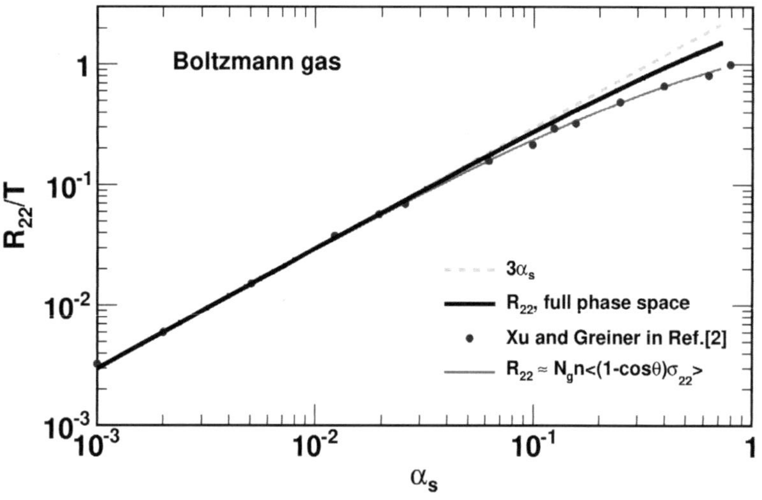

Figure 1. The elastic collision rate R_{22} in a Boltzmann gas.

i.e. we take full phase space including both channels, then the factor of 2 is not needed:

$$|M_{12\to 34}|^2_{CM} \underset{\mathbf{q}_T^2\approx 0}{\approx} 72\pi^2\alpha_s^2\frac{s^2}{(\mathbf{q}_T^2+m_D^2)^2}. \tag{2}$$

Note that the constraint $\mathbf{q}^2 \approx 0$ is removed because both the near forward and backward scatterings have small $\mathbf{q}_T^2$ but only the near forward scatterings have small $\mathbf{q}^2$ [8].

With the matrix element, it is straightforward to calculate the elastic collision rate. Using the small $\mathbf{q}$ approximation, the integral of $d^3p_3d^3p_4$ can be simplified as

$$\int \frac{d^3p_3}{(2\pi)^3}\frac{d^3p_4}{(2\pi)^3}(2\pi)^4\delta^4(p_1+p_2-p_3-p_4)[\cdots] \approx \frac{1}{4\pi^2}\int d^3q\delta(\hat{\mathbf{k}}_1\cdot\mathbf{q}-\hat{\mathbf{k}}_2\cdot\mathbf{q})[\cdots]. \tag{3}$$

The δ function can be removed by the integral over $\mathbf{q}$, further integral over $\mathbf{p}_1$ and $\mathbf{p}_2$ gives the result $R_{22} = \frac{\pi^2}{2\zeta(3)}\alpha_s T \approx 4.105\alpha_s T$ for a boson gas with $m_D^2 = 4\pi\alpha_s T^2$. For a Boltzmann gas with $m_D^2 = 24/\pi\alpha_s T^2$, the Bose enhancement factor $(1+f_3^0)(1+f_4^0)$ is absent, and the result is $R_{22} = 3\alpha_s T$. To check the above analytic results, we can carry out the collisional integral numerically using the matrix element in Eq.(2) in full phase space. The result is shown in Fig.1 by the thick solid line, which matches the dashed line of $R_{22} = 3\alpha_s T$ at $\alpha_s < 0.02$ perfectly. The difference between the two curves for large α_s is due to the fact that the small $\mathbf{q}$ approximation doesn't hold for large m_D.

Another way of calculating the rate is through the cross section [9] with which the elastic collision rate in a Boltzmann gas can be written as,

$$R_{22} = \frac{N_g}{n}\int \frac{d^3p_1}{(2\pi)^3}\frac{d^3p_1}{(2\pi)^3}\frac{p_1\cdot p_2}{E_1E_2}f_1^0f_2^0\sigma_{22} = N_g n < (1-\cos\theta_{12})\sigma_{22} > . \tag{4}$$

If neglecting the s dependence in σ_{22}, it is easy to get $R_{22} = 3\alpha_s T$. With the s dependence, the average over the $\mathbf{p}_1$ and $\mathbf{p}_2$ gives the thin solid line in Fig.1, which agrees with the Xu and Greiner's result [2] quite well.

For the 23 process, the exact tree level matrix element [10, 11, 12] is given, e.g., in Eqs. (5,6,12) of Ref. [8]. In the limit of one particular configuration in phase space, namely,

p_1, p_2, p_3, p_4 are hard (their momenta are $O(\sqrt{s})$), while $q(= p_2 - p_4 = p_3 + p_5 - p_1)$ and $k = p_5$ are soft (momenta are much smaller than $\sqrt{s}$), the matrix element becomes

$$|M_{12\to 345}|^2_{CM} \approx 3456\pi^3\alpha_s^3 \frac{s^2}{(k_T^2 + m_g^2)\,(q_T^2 + m_D^2)\left[(\mathbf{k}_T - \mathbf{q}_T)^2 + m_D^2\right]}. \quad (5)$$

It can be seen that the Gunion-Bertsch (GB) formula [13] is recovered after taking $m_D, m_g \to 0$, and in this limit, the denominator can be written in a symmetric form $\mathbf{p}_{T3}^2\mathbf{p}_{T4}^2\mathbf{p}_{T5}^2$. There are 6 different configurations in the full phase space, represented by the permutation of $\{3, 4, 5\}$. If the phase space of a single configuration can be isolated and integrated in collisional integrals, the result times 6 should be equal to the full phase space result. We will use the above configuration where k_5 is soft to calculate the rate of the 23 process with the symmetry factor. We now take some approximations as follows: (1) For small k, the Bose enhancement diverges as $1+f_k \sim 1/k$. (2) As the contribution from small k is enhanced by the divergence, we may assume $k \ll q$ and $|\mathbf{k}_T| \ll |\mathbf{q}_T|$, then the integrals over k and q are decoupled. (3) The recoil from soft gluon to hard gluons is negligible, we have $\delta^4(p_1 + p_2 - p_3 - p_4 - p_5) \approx \delta^4(p_1 + p_2 - p_3 - p_4)$, and Eq.(3) is still valid. So the inelastic collision rate can be written as

$$\begin{aligned} R_{23} &= 6 \times \frac{N_g 3456\pi^3\alpha_s^3}{6n} \int p_1^2 dp_1 (2\pi)\sin(\theta_{12})d\theta_{12} \int p_2^2 dp_2 (4\pi) \\ &\times \int k^2 dk \sin(\theta_k)d\theta_k d\phi_k \int q^2 dq \sin(\theta_q)d\theta_q d\phi_q \\ &\times \frac{\delta(\phi_q - \theta_{12}/2)}{q\sin\theta_q \sin(\theta_{12}/2)} \frac{f_1 f_2(1+f_1)(1+f_2)(1+f_k)}{(2\pi)^{11}2^5 E_1^2 E_2^2 k} \\ &\times \frac{s^2}{(\mathbf{q}_T^2 + m_D^2)^2} \cdot \frac{\mathbf{q}_T^2}{(\mathbf{k}_T^2 + m_D^2)(\mathbf{q_T}^2 + m_D^2)}. \end{aligned} \quad (6)$$

Note that symmetry factor 6 is added, and the GB formula for the matrix element in literature is used. Although complicated enough, most of the integral can be integrated out. We will end up with three entangled angular integrals which have to be computed numerically. The final result gives $R_{23} = 4.818\alpha_s^{3/2}$. Note that the divergence of Bose distribution in soft regime plays a very important role. In a Boltzmann gas, these approximations do not hold and the analytical approach do not work. In a general case, we should integrate over full phase space without the symmetry factor by solving $\delta^4(p_1 + p_2 - p_3 - p_4 - p_5)$ exactly and numerically. The results from two methods are shown as the black solid line and red dashed-dotted line in Fig.2. The sizable difference at large α_s comes from the fact that the small $\mathbf{q}_T$ approximation does not hold for large m_D.

Another way to regulate the IR singularity is to use the Landau-Pomeranchuk-Migdal (LPM) effect [14, 15] to provide a cutoff for the bremsstrahlung gluon. One can simply use $\Theta(\mathbf{k}_T^2 - E_k/l_{mfp})$ in the collisional integral, where the the mean free path can be estimated by the inverse of the collision rate $l_{mfp} \sim 1/R = 1/(R_{22} + 2.5R_{23})$. The semi-analytic method is still workable and gives the self-consistent equation $R_{23}^{LPM} = 6.52\alpha_s^2/(R_{22} + 2.5R_{23}^{LPM})$, whose solution is $R_{23}^{LPM} = \frac{1}{5}(\sqrt{R_{22}^2 + 65.2\alpha_s^2} - R_{22}) \approx 0.99\alpha_s T$. We can also get a self-consistent result by iterating the numerical integration over the full phase space, in which the LPM condition applied to the softest gluon for each configuration satisfying the energy-momentum conservation. The results from the two methods agree with each other perfectly at small α_s as shown by the green solid and black long-dashed line in Fig.2.

In summary, we have calculated elastic and inelastic collision rates in a gluon gas with different methods to show the significance of the symmetry factor in phase space integrals. The current

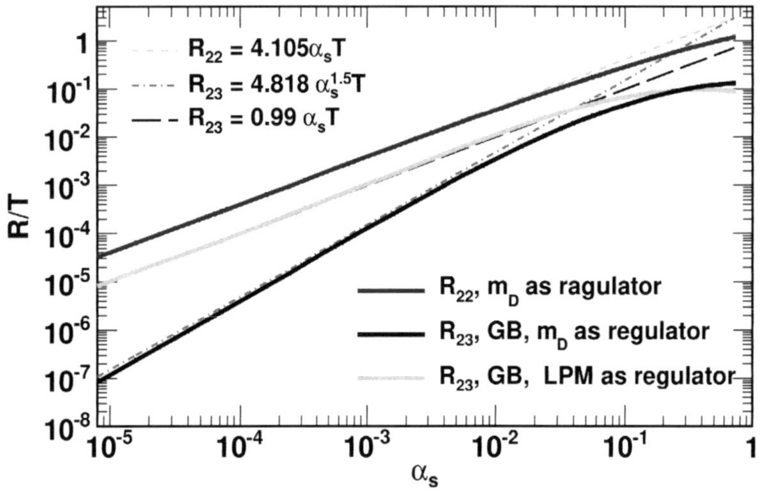

Figure 2. The elastic and inelastic collision rate for a boson gas. Solid lines stand for the numerical results in full phase space without the symmetry factor, long-dashed, short-dashed and dashed-dotted line stand for semi-analytic results in the constrained phase space with the symmetry factor.

study confirms our previous result about the role of the symmetry factor in the shear and bulk viscosity [6, 7, 8]. We have illustrated the equivalence of carrying out collision integral in the constrained phase space with the symmetry factor or in full phase space without the symmetry factor. Our result shows that the 22 process is more important than 23 process with reasonable coupling constant where the perturbative QCD is reliable. With larger coupling constant, for example, in the QGP created at the RHIC/LHC energies, $\alpha_s \sim 0.2 - 0.3$, non-perturbative effects should be taken into account to explain the observed small shear viscosity $\eta/s \sim 1/(4\pi)$. The current scheme can be applied to calculate some other thansport coefficients, such as the dragging coefficient of heavy quarks, the jet quenching parameter, etc., on which we will look at in the future.

QW is supported in part by the National Natural Science Foundation of China under grant 11125524. JD is supported by the National Natural Science Foundation of China under grant 11105082.

[1] P. Kovtun, D. T. Son and A. O. Starinets, Phys. Rev. Lett. **94**, 111601 (2005).
[2] Z. Xu and C. Greiner, Phys. Rev. Lett. **100**, 172301 (2008).
[3] C. Wesp, A. El, F. Reining, Z. Xu, I. Bouras and C. Greiner, Phys. Rev. C **84**, 054911 (2011).
[4] P. Arnold, G. D. Moore and L. G. Yaffe, JHEP **0011**, 001 (2000).
[5] P. Arnold, G. D. Moore and L. G. Yaffe, JHEP **0305**, 051 (2003).
[6] J. W. Chen, H. Dong, K. Ohnishi and Q. Wang, Phys. Lett. B **685**, 277 (2010).
[7] J. W. Chen, J. Deng, H. Dong and Q. Wang, Phys. Rev. D **83**, 034031 (2011) [Erratum-ibid. D **84**, 039902 (2011)].
[8] J. W. Chen, J. Deng, H. Dong and Q. Wang, arXiv:1107.0522 [hep-ph].
[9] Z. Xu and C. Greiner, Phys. Rev. C **76**, 024911 (2007) [hep-ph/0703233].
[10] F. A. Berends, R. Kleiss, P. De Causmaecker, R. Gastmans and T. T. Wu, Phys. Lett. B **103**, 124 (1981).
[11] R. K. Ellis and J. C. Sexton, Nucl. Phys. B **269**, 445 (1986).
[12] T. Gottschalk and D. W. Sivers, Phys. Rev. D **21**, 102 (1980).
[13] J. F. Gunion and G. Bertsch, Phys. Rev. D **25**, 746 (1982).
[14] M. Gyulassy, M. Plumer, M. Thoma and X. N. Wang, Nucl. Phys. A **538**, 37C (1992).
[15] X. N. Wang, M. Gyulassy and M. Plumer, Phys. Rev. D **51**, 3436 (1995).

Journal of Physics: Conference Series **422** (2013) 012008 doi:10.1088/1742-6596/422/1/012008

Quarkonium suppression from parton energy loss

Stéphane Peigné

SUBATECH, UMR 6457, Université de Nantes, Ecole des Mines de Nantes, IN2P3/CNRS
4 rue Alfred Kastler, 44307 Nantes cedex 3, France

E-mail: `peigne@subatech.in2p3.fr`

Abstract. The effects of energy loss in cold nuclear matter on J/ψ suppression in p–A collisions are studied. A simple model based on first principles and depending on a single free parameter is able to reproduce J/ψ suppression data at large x_F or large rapidity and at various center of mass energies. These results strongly support parton energy loss as a dominant effect in quarkonium suppression. Predictions for J/ψ and Υ suppression in p–Pb collisions at the LHC are made.

1. Introduction

Recently, new scaling properties have been identified for the induced gluon radiation spectrum $dI/d\omega$, and associated energy loss ΔE, of certain hard processes [1]. The phenomenological consequences of these results on J/ψ and Υ nuclear suppression in p–A and π–A collisions have then been studied [2]. In this talk I present the main idea and results of Refs. [1, 2].

I will show that the large x_F quarkonium suppression data in p–A collisions, parametrized by

$$R^{\psi}_{\mathrm{pA}}\left(x_F, \sqrt{s}\right) = \frac{1}{A}\frac{d\sigma^{\psi}_{\mathrm{pA}}}{dx_F}\left(x_F, \sqrt{s}\right) \Big/ \frac{d\sigma^{\psi}_{\mathrm{pp}}}{dx_F}\left(x_F, \sqrt{s}\right), \tag{1}$$

can be explained by parton energy loss in cold matter. Although $\hat{q}$ in cold matter is small, a strong nuclear attenuation arises due to the specific parametric behaviour $\Delta E \propto E$ at sufficiently large E, where E is the quarkonium energy in the target nucleus rest frame. As discussed in Ref. [1], this behaviour holds when the partonic subprocess can be viewed (in the nucleus rest frame) as the small angle scattering of a color charge. In the following we focus on quarkonium hadroproduction, where the mass M of the heavy-quark $Q\bar{Q}$ pair provides the hard scale.

The physics of J/ψ nuclear suppression depends crucially on the precise kinematics under consideration. When the J/ψ hadronization time $t_{hadro} \ll L$, with L the nuclear size, nuclear suppression

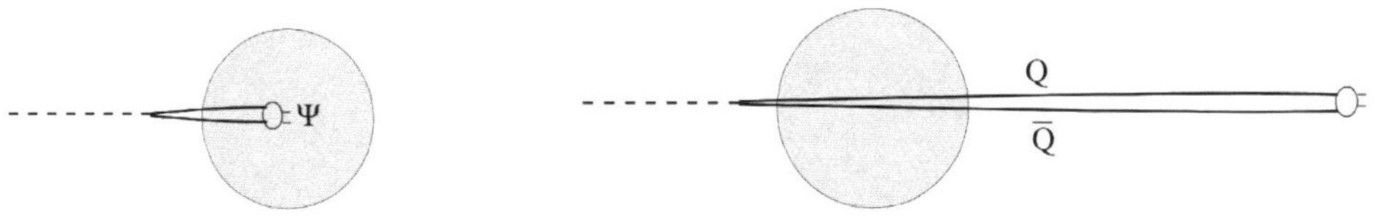

Figure 1. Quarkonium hadroproduction at small (left) and large (right) quarkonium energy E in the nucleus rest frame. In the latter case a compact color octet $Q\bar{Q}$ pair crosses the nucleus.

arises from the absorption cross section σ_{abs} of the fully formed J/ψ in the nucleus (see Fig. 1 left). When $t_{hadro} \gg L$, what propagates through the nucleus is a compact color octet $c\bar{c}$ pair (Fig. 1 right), and σ_{abs} becomes irrelevant. In the following we focus on the latter kinematical domain. In the Figures presented in Section 4, we indicate by an arrow the typical value of x_F (or rapidity) where $t_{hadro} = L$, above which our model should be valid.

2. Parton energy loss in J/ψ hadroproduction

The presence of a large (medium-induced) parton energy loss associated to J/ψ hadroproduction can be understood as follows. At large x_F, the gluon emission required for the $c\bar{c}$ color neutralization is constrained to be softish by energy conservation, and thus the $c\bar{c}$ pair remains color octet for a long time $t_{octet} \gg t_{hard}$, where $t_{hard} \sim E/M^2$ is the perturbative production time of the $Q\bar{Q}$ pair. As a consequence, at large x_F the J/ψ *hadro*production subprocess looks like small angle scattering of a color charge, as pictured in Fig. 2. The associated radiation spectrum arising from gluon formation times t_f satisfying $t_{hard} \ll t_f \ll t_{octet}$ is easy to calculate [1]. It is similar to the (non-abelian) Bethe-Heitler spectrum of an asymptotic charge, and depends on the amount of transverse momentum kick $q_\perp$ to the charge. The typical $q_\perp$ is expected to be larger in p-A than in p-p collisions due to transverse momentum nuclear broadening $\Delta q_\perp^2 = \hat{q}L$. The resulting medium-induced radiation spectrum is similar to the Bethe-Heitler spectrum with $q_\perp^2$ replaced by $\Delta q_\perp^2$ [1], and the average energy loss scales as the quarkonium energy, $\Delta E \propto E$.

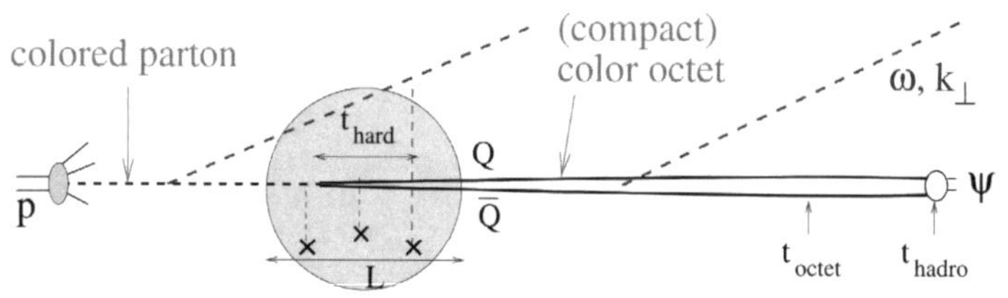

Figure 2. Generic process for J/ψ production in p-A collisions, assuming $L, t_{hard} \ll t_{octet} \lesssim t_{hadro}$. An induced radiated gluon $(\omega, \mathbf{k}_\perp)$ with formation time t_f satisfying $L, t_{hard} \ll t_f \ll t_{octet}$ is represented.

The energy dependence $\Delta E \propto E$ for J/ψ hadroproduction originates from coherence between initial and final state radiation. We expect $\Delta E \propto E$ in all situations where an incoming (color) charge is scattered to an outgoing charge, as in open charm and light hadron production, see Fig. 3 (top). However $\Delta E \propto E$ is not expected in DIS nor Drell-Yan production off nuclei (Fig. 3 bottom), where the purely final-state (initial-state) DGLAP-like radiation (with associated $t_f \gg t_{hard}$) cancels in the medium-induced spectrum.

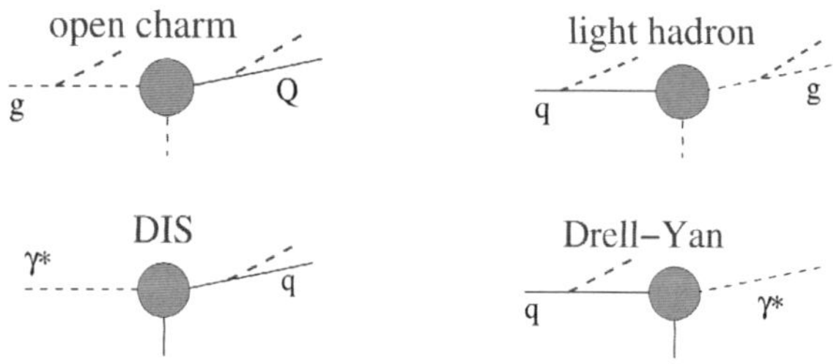

Figure 3. Partonic subprocesses where a medium-induced loss $\Delta E \propto E$ arising from coherent radiation is expected (top) or not (bottom).

Journal of Physics: Conference Series **422** (2013) 012008 doi:10.1088/1742-6596/422/1/012008

3. Simple model for J/ψ nuclear suppression

Our model consists in expressing the J/ψ differential production cross section $\mathrm{d}\sigma/\mathrm{d}x_F$ in p–A collisions simply as that in p–p collisions, with a shift in x_F accounting for the energy loss ε incurred by the octet $c\bar{c}$ pair propagating through the nucleus,

$$\frac{1}{A}\frac{\mathrm{d}\sigma_{\mathrm{pA}}^{J/\psi}}{\mathrm{d}x_F}(x_F) = \int_0^{\varepsilon_{max}} \mathrm{d}\varepsilon\, \mathscr{P}(\varepsilon)\, \frac{\mathrm{d}\sigma_{\mathrm{pp}}^{J/\psi}}{\mathrm{d}x_F}(x_F + \delta x_F(\varepsilon)), \tag{2}$$

where $x_F(E) + \delta x_F(\varepsilon) = x_F(E+\varepsilon)$. At large enough x_F we have $E \simeq x_F E_{\mathrm{p}}$ and $\delta x_F(\varepsilon) \simeq \varepsilon/E_{\mathrm{p}}$, where E_{p} is the incoming proton energy. The other ingredients in Eq. (2) are:
(i) The differential p–p cross section, determined from a fit of p–p data;
(ii) The energy loss probability distribution, or *quenching weight* $\mathscr{P}(\varepsilon)$, directly related to the medium-induced radiation spectrum $\mathrm{d}I/\mathrm{d}\omega$ associated to the hard subprocess (see Fig. 2) and derived in [1]. Since this spectrum is fully determined by $\Delta q_\perp^2 = \hat{q}L$, the model depends on the single parameter $\hat{q}$.

In fact, $\hat{q}$ is related to the gluon distribution $G(x)$ in a target nucleon as $\hat{q}(x) \propto xG(x)$ [3]. Since $xG(x) \sim x^{-0.3}$ is suggested by fits to HERA data [4], $\hat{q}$ actually depends on $x = x_2$, where x_2 is the target parton momentum fraction, $\hat{q}(x) = \hat{q}_0(0.01/x)^{0.3}$. Note that the x-dependence of $\hat{q}$ is not essential. Similar final results would be obtained using a constant $\hat{q}(x) = \hat{q}$. In the following the single parameter of the model is taken as $\hat{q}_0 = \hat{q}(x = 0.01)$. The model defined by Eq. (2) allows to predict the effects of parton energy loss on the nuclear attenuation factor (1).

Besides energy loss effects, in the high energy domain under consideration we expect saturation of the nuclear gluon density at small x_2, leading to an additional J/ψ suppression mechanism in p–A collisions. In order to make reliable predictions at RHIC and LHC, where saturation effects might be important, the J/ψ nuclear production ratio has been determined assuming energy loss effects, with and without saturation effects. (For the implementation of saturation in the model, see Ref. [2].)

4. Results

The parameter $\hat{q}_0$ is determined by the best fit provided by the model to the E866 fixed target data [5] for the ratio $R_{\mathrm{W/Be}} \equiv R_{\mathrm{pW}}/R_{\mathrm{pBe}}$, see Fig. 4 (left). The fit gives $\hat{q}_0 = 0.09$ GeV^2/fm assuming energy loss effects only, and $\hat{q}_0 = 0.05$ GeV^2/fm when saturation effects are also included. The successful description of J/ψ suppression in iron, $R_{\mathrm{Fe/Be}}$ (Fig. 4 right), at the same energy fully supports the atomic mass dependence of the model. Note that phase space limitation effects are included in the model, due to the use of $\mathscr{P}(\varepsilon)$ in (2) and $\varepsilon_{max} \simeq (1-x_F)E_{\mathrm{p}} \to 0$ when $x_F \to 1$. The disagreement with the E866 data

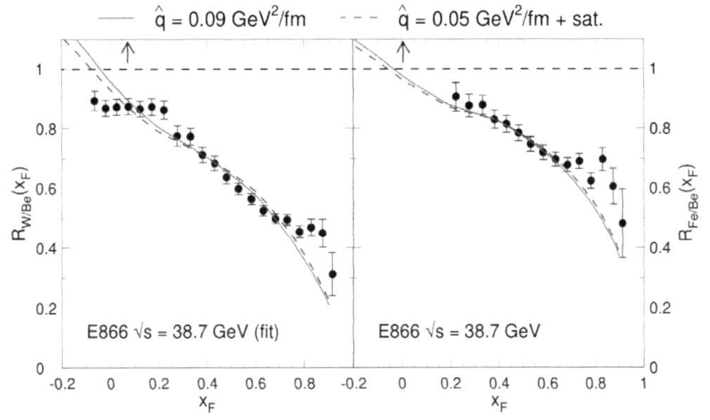

Figure 4. E866 J/ψ suppression data [5] compared to the energy loss model.

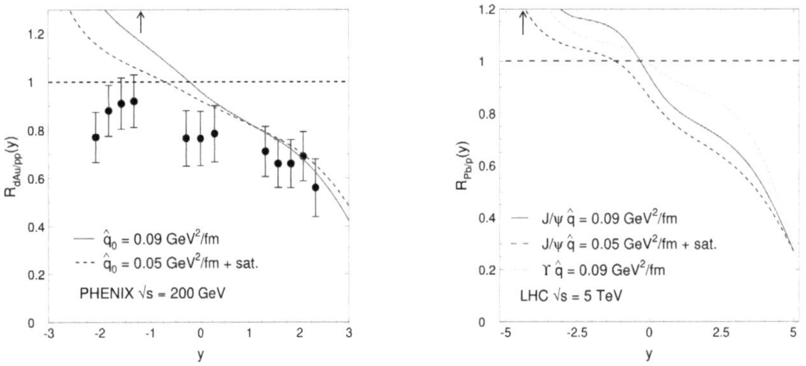

Figure 5. Left: PHENIX J/ψ suppression data [9] in d–Au collisions compared to the energy loss model; Right: J/ψ and Υ suppression expected in p–Pb collisions at the LHC.

at the highest x_F values might be due to our assumption of an incoming *gluon*, which becomes invalid at very large x_F where *quark*-induced processes come into play and might lead to a smaller energy loss.

In Ref. [2], lower $\sqrt{s}$ (and smaller x_F) J/ψ suppression data in p–A collisions (from NA3 [6], NA60 [7] and HERA-B [8]) are compared to the energy loss model, with the same value of $\hat{q}_0$. The agreement is very good over a very wide range in x_F [2]. The predictions including saturation are very similar to those without. As expected saturation effects are tiny at these energies.

The predictions in d–Au collisions at RHIC, $\sqrt{s} = 200$ GeV, are shown in Fig. 5 (left) in comparison with PHENIX data [9]. Energy loss effects are able to reproduce J/ψ suppression at positive rapidities. Not surprisingly, some disagreement is observed in the negative y bins for which nuclear absorption should play a role. Finally, arguing about the possible slight disagreement observed around mid-rapidity might be premature, given the present experimental uncertainties.

Last, the rapidity dependence of J/ψ suppression in p–Pb collisions at the LHC (taking $\sqrt{s} = 5$ TeV) is shown in Fig. 5 (right). Even though the inclusion of saturation effects is expected to yield a stronger J/ψ suppression, it is somehow compensated by the use of a smaller transport coefficient; as a consequence J/ψ predictions with (dashed line) and without (solid line) saturation are actually rather similar, except in the negative y-bins. Fig. 5 (right) also shows the predicted Υ suppression as a dash-dotted line. Because of the mass dependence of energy loss, $\Delta E \propto 1/M$ [1, 2], it is expected to be smaller than that of J/ψ yet not negligible.

In summary, an energy loss model (supplemented by saturation effects) based on first principles has been presented. Once the transport coefficient $\hat{q}$ is determined from E866 data, the model is able to reproduce nicely all existing J/ψ measurements in p–A collisions. The studies of Refs. [1, 2] thus support parton energy loss as a dominant effect in p-A J/ψ suppression. The model predictions for J/ψ and Υ nuclear suppression at LHC energies can be compared to the future measurements by the ALICE and LHCb experiments during the p–Pb run scheduled in January 2013.

[1] F. Arleo, S. Peigné and T. Sami, Phys. Rev. D **83**, 114036 (2011) [arXiv:1006.0818 [hep-ph]].
[2] F. Arleo and S. Peigné, Phys. Rev. Lett. **109** 122301 (2012) [arXiv:1204.4609 [hep-ph]].
[3] R. Baier, Y. L. Dokshitzer, A. H. Mueller, S. Peigné and D. Schiff, Nucl. Phys. B **484**, 265 (1997) [arXiv:hep-ph/9608322].
[4] K. J. Golec-Biernat and M. Wusthoff, Phys. Rev. D **59**, 014017 (1998) [arXiv:hep-ph/9807513].
[5] M. J. Leitch *et al.* [FNAL E866/NuSea collaboration], Phys. Rev. Lett. **84**, 3256 (2000) [arXiv:nucl-ex/9909007].
[6] J. Badier *et al.* [NA3 Collaboration], Z. Phys. C **20**, 101 (1983).
[7] R. Arnaldi *et al.* [NA60 Collaboration], Phys. Lett. B **706**, 263 (2012) [arXiv:1004.5523 [nucl-ex]].
[8] I. Abt *et al.* [HERA-B Collaboration], Eur. Phys. J. C **60**, 525 (2009) [arXiv:0812.0734 [hep-ex]].
[9] A. Adare *et al.* [PHENIX Collaboration], Phys. Rev. Lett. **107**, 142301 (2011) [arXiv:1010.1246 [nucl-ex]].

Journal of Physics: Conference Series **422** (2013) 012009 doi:10.1088/1742-6596/422/1/012009

Dynamical Quarkonia Suppression from Complex Potentials

J. Casalderrey-Solana

Departament d'Estructura i Constituents de la Matèria and Institut de Ciències del Cosmos (ICCUB), Universitat de Barcelona, Martí i Franquès 1, 08028 Barcelona, Spain

E-mail: jorge.casalderrey@ub.edu

Abstract. I address the effect that a temperature dependent potential has on the suppression of heavy quarkonia states in deconfined hadronic matter. I focus on a simple medium: a homogenous, fixed temperature and deconfined system with a finite lifetime (QGP-brick). Assuming that all the interactions of a heavy quark anti-quark ($Q - \bar{Q}$) pair with the medium can be recast into an in-medium potential, I solve the time dependent Schrödinger equation to evolve the density matrix which describes the hard pair production and its connection to the final distribution of hadrons after the medium disappears. For those temperatures in which bound states survive in the medium, I find a non-trivial dependence of the production of excited quarkonia states on the in-medium levels, due to the mixing of vacuum and in-medium wave functions. This mixing leads, in particular, to the enhancement of the relative abundance of 2S to 1S states for the case in which the in-medium ground state of the $Q - \bar{Q}$ system is dissolved.

1. Introduction

The physics of quarkonia suppression in heavy ion collisions has experienced a radical change in the last few years. On the one hand, experimental advances at the LHC (and also at RHIC) have opened access to the bottomonium family. On the theory side, the survival of quarkonia above deconfinement together with the realization that the static properties of in-medium bound states must include complex heavy quark potentials have radically changed the picture of quarkonia dynamics in a deconfined plasma. In this proceedings I will explore the effects that these two features have on the pattern of production of the different quarkonia states.

Following [1], I will study quarkonia production in a homogeneous medium at fixed temperature with a finite lifetime after which the temperature drops abruptly to zero. The spacial extent of the medium is taken to be large, much larger that the in-medium radius of bound states. I will assume that a $Q - \bar{Q}$ pair is formed within the medium via a hard process at a time $t_0 = 0$ and interacts with it during a finite time t after which the medium disappears. After this time, the evolution occurs as in vacuum. In analogy with the jet quenching studies, where a similar set-up was used to analyze the differences among the available quenching models, I will call this set-up, which I have sketched in fig. (1), a QGP-brick.

2. Formalism

From the point of view of the soft dynamics, the hard production of the $Q - \bar{Q}$ can be described via a Wigner function, $W_H(\mathbf{r}, \mathbf{q})$. Independently of the particular process that leads to the

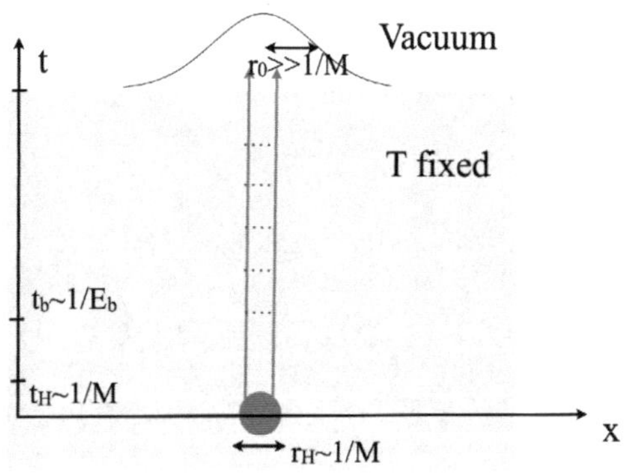

Figure 1. Sketch of the QGP-Brick set up. At initial time a $Q-\bar{Q}$ pair is formed in the medium via a hard process that occurs within a short time $t_H \sim 1/M$. At a later time of order the inverse binding energy, the soft interactions lead to the formation of quarkonia states from the pair, which are influenced by the presence of the medium. At a later time t, the medium disappears and the $Q-\bar{Q}$ system projects into vacuum quarkonia states.

production of a $Q-\bar{Q}$ pair within the medium, since the process is hard, the typical momentum of each of the quarks, p, is large $M < p$. Thus, the typical relative quark momentum is also much larger than the inverse bound state radius and, from the point of view of the soft matrix element, only the $q \to 0$ limit of W_H is relevant for the production of bound states, which after a Fourier transform, leads to an approximate initial density matrix given by

$$\rho_H\left(\mathbf{r}-\frac{\mathbf{y}}{2},\mathbf{r}+\frac{\mathbf{y}}{2};t_0\right)=\delta(\mathbf{r})\delta\left(\mathbf{y}\right), \tag{1}$$

where I have assumed that the hard process takes place at very small distances.

As the $Q-\bar{Q}$ pair propagates, the density matrix eq (1) evolves via a non-relativistic hamiltonian, since the late time interactions are soft as compared to the pair mass. At any given time $t > t_0$ the different particle yields can be obtained by projecting the evolved density matrix into the single state density matrix

$$Y_S \propto \mathrm{Tr}\left(\rho_H(t)\rho_s\right), \tag{2}$$

with $\rho_s = |S\rangle\langle S|$ and $|S\rangle$ the quarkonium estate of interest. Using the cyclicity of the trace, the forward time-evolution of the initial density matrix can be transformed to a backwards evolution of the final quarkonium wave function. Defining $\tilde{t} = t - t_0$ this wave function satisfies the (hermitian conjugate) Schrödinger equation:

$$-i\partial_{\tilde{t}}\tilde{\psi}(\tilde{t},\mathrm{r}) = -\frac{\nabla^2}{M_q}\tilde{\psi}(\tilde{t},\mathrm{r}) + V^{\dagger}(r)\tilde{\psi}(\tilde{t},\mathrm{r}), \tag{3}$$

with $\tilde{\psi}(\tilde{t},\mathrm{r}) = \left\langle \mathbf{r}|\tilde{S}\right\rangle$ and initial condition $\tilde{\psi}(0,\mathrm{r}) = \langle \mathbf{r}|S\rangle$. Combining eq. (1) and eq. (2), the yield of quarkonia production after a time of propagation t is given by

$$Y_S \propto \left|\tilde{\psi}(t-t_0,\mathbf{r}=0)\right|^2, \tag{4}$$

which, in the vacuum, coincides with the relevant long distance matrix element for production in the singlet model. For production in the medium-brick, the interaction with the QGP fields

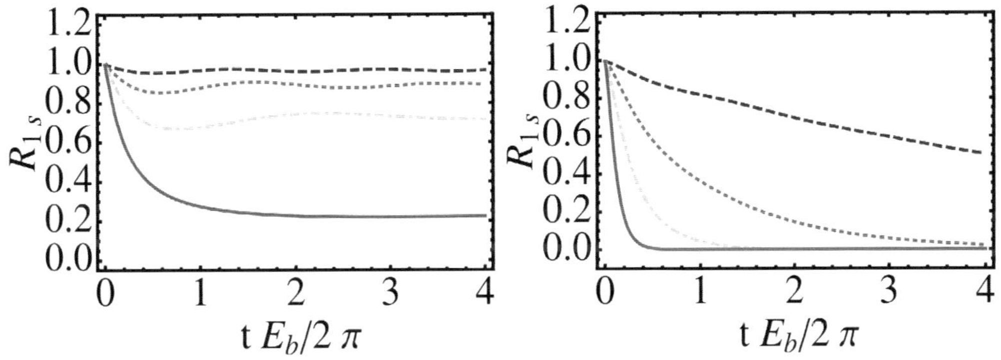

Figure 2. Yield suppression of the ground state as a function of the lifetime of the QGP in units of the vacuum ground state period $\mathcal{T} = 2\pi/E_b$ for real (left) and complex potentials (right) for $2\mu_D/\alpha M = 0.15,\ 0.3,\ 0.5,\ 0.9$ for the dashed, dotted, dash dotted and solid lines respectively

changes the potential during the time extent of the medium and changes abruptly back to the vacuum expression when the $Q - \bar{Q}$ leaves the QGP-brick. The complicated dynamics of the production process are given by the time dependent Schrödinger equation, eq. (3). In this study I will focus in a complex potential of the form.

$$V(r) = -\alpha \left(\frac{e^{-\mu_D r}}{r} + i\mu_D\, \phi(\mu_D r) \right), \tag{5}$$

where μ_D is the Debye screening length of the plasma and

$$\phi(r) = \int_0^\infty dz \frac{z}{(z^2+1)^2} \left(1 - \frac{\sin(zr)}{zr}\right), \tag{6}$$

which coincides with the functional form of the imaginary part of the perturbative potential [2, 3]. To understand the effect of this imaginary part on the yield, I will compare the results of this potential with a real potential given by $\mathrm{Re}\, V(r)$.

Since after the particle leaves the brick the yield of the different states remains constant, the modification of the quarkonia yields is controlled by the evolution of the hard density matrix up to the time t when the medium disappears. The yield suppression is, then, given by ratio of yields in medium and in vacuum at this time.

$$R_S = \frac{Y_S(\mu_D)}{Y_S(\mu_D = 0)}. \tag{7}$$

3. Results and Conclusions

In fig. (2) I show the suppression of the S-wave ground state (1S) for different values of μ_D as a function of the time duration of the box for both the real and complex potentials. For very small brick lifetimes, less than one period, the suppression is relatively modest for all values of μ_D. For the real potential, the late time suppression saturates to a constant value as long as the in medium potential supports a bounded ground state. When μ_D is so large that the ground state disappears the late time yield tends to zero. The complex potential, eq. (5), is much more effective in suppressing quarkonia. In this case the late time suppression does not saturate to a constant, but continuously increases due to the finite width of the in-medium states. In fact, at

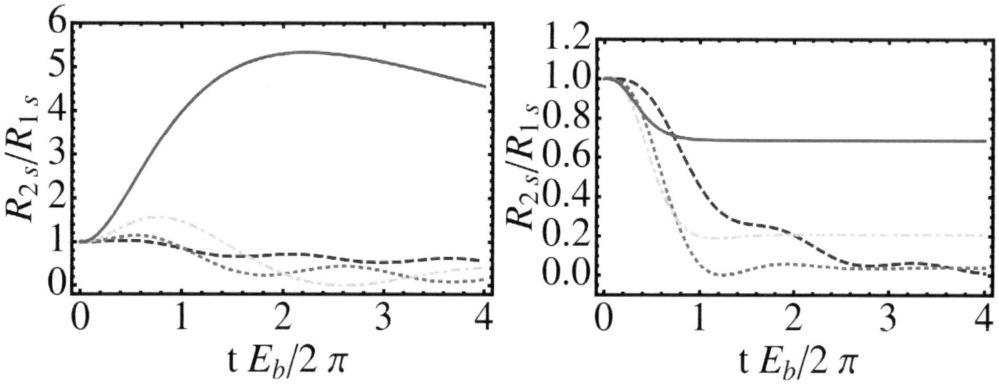

Figure 3. Double ratio of the suppression factors of 2S and 1S states as function of the lifetime of the QGP in units of the vacuum ground state period $\mathcal{T} = 2\pi/E_b$ for real (left) and complex potentials (right) for $2\mu_D/\alpha M = 0.15, 0.3, 0.5, 0.9$ for the dashed, dotted, dash dotted and solid lines respectively

these late times, the dynamics of the in-medium ground state is dominated by this absorption. It can be checked that after the formation time of the meson the absorption is indeed given by the in-medium width up to an overall shift, which is due to the early time suppression.

In Fig (3) I show the double ratio of the suppression factor of the 2S state to the 1S estate for both potentials. For the real potential, as the temperature rises a modulation in the double yield ratio appears which is due to the non-zero overlap of the vacuum 2S state with the in-medium ground state. This double ratio can be larger than 1, showing a relative enhancement of 2S to 1S states. This is particularly dramatic in the highest temperature case, for which the bound states are loosely bound. This behavior continues at even higher temperatures, when all states are dissolved and it is due to the fact that the vacuum 2S state is wider than the ground state and has a larger overlap with the in-medium continuum. For the complex potential, the late time ratio is constant, which means that both the 2S and 1S yields fall at the same rate. Similarly to the real potential case, this is due to the fact that when the in-medium 2S level is absent, the production is dominated by the overlap with the in-medium 1S. As in the real potential case, the asymptotic value grows with μ_D and becomes greater than one at larger temperatures.

In summary, by comparing the suppression patterns of a medium-brick characterized by a real quarkonia potential to that of a potential with an imaginary part with the same screening parameter μ_D I have shown that the non-unitarity effects introduced by the complex potential are very efficient in suppressing quarkonia. Additionally, I have shown that even though the ground state suppression pattern is easy to understand in terms of in-medium bound states, the suppression of excited states, such as 2S-wave level, is due to the mixing with the in-medium 1S state. Such mixing leads to an enhancement of the double ratio of suppression patterns of 2S to 1S states which is due to the wider wave function of the in-medium states. Since this behavior is generic, this type of effects may be behind the anomalous suppression pattern of $\psi(2S)$ states reported by the CMS collaboration [4].

Acknowledgments: I am supported by a Ramón y Cajal fellowship. I also acknowledge financial support by the research grants FPA2010-20807, 2009SGR502 and by the Consolider CPAN project..

[1] J. Casalderrey-Solana, arXiv:1208.2602 [hep-ph].
[2] M. Laine, O. Philipsen, P. Romatschke and M. Tassler, JHEP **0703**, 054 (2007) [hep-ph/0611300].
[3] A. Beraudo, J. -P. Blaizot and C. Ratti, Nucl. Phys. A **806**, 312 (2008) [arXiv:0712.4394 [nucl-th]].
[4] CMS Collaboration Collaboration, S. Chatrchyan et al.,CMS-PAS-HIN-12-007

Journal of Physics: Conference Series **422** (2013) 012010 doi:10.1088/1742-6596/422/1/012010

Probing the medium with jets in ALICE

Pasquale Di Nezza for the ALICE Collaboration

Laboratori Nazionali di Frascati, INFN,
via E.Fermi 40, I-00044, Frascati (Rome) - Italy

E-mail: Pasquale.DiNezza@lnf.infn.it

Abstract. Among the different hard probes, jet results have been here selected to present a unique insight into the nature of the hot QCD matter. Jets are reconstructed from charged particles using the anti-k_T algorithm with the ALICE detector at the LHC in Pb-Pb collisions at $\sqrt{s_{NN}} = 2.76$ TeV in a momentum range down to 0.150 GeV and in a pseudorapidity range $|\eta| < 0.9$. We report inclusive jet spectra and ratios as a function of the jet transverse momenta (p_T) for different centrality classes and jet radii. Strong suppression of the inclusive jet spectrum has been observed in the most central collisions. Comparison with theory puts more stringent constraints on modeling jet modification and energy loss in dense matter.

1. Introduction

In the study of the Quark Gluon Plasma (QGP), the hard probes are commonly considered the direct probes able to resolve the sub-hadronic scales and distinguish confined and deconfined media in the different steps of the fragmentation. Jets with large transverse momentum play an essential role in the space-time tomography study of the relativistic heavy-ion collisions and of the QGP. This is because, being produced in the early stage of the reaction, prior to the formation of the hot medium, jets interact with the traversed dense matter through elastic and inelastic collisions loosing some of their initial energy, quenching [1], in the process.

In this proceeding, ALICE preliminary results on charged jet reconstruction, in Pb-Pb collisions at $\sqrt{s_{NN}} = 2.76$ TeV, are discussed. Single inclusive charged jet cross sections for different radii are presented. Jet suppression and jet broadening are explored.

The jet momenta contamination from soft production and the remaining influence of underlying event fluctuations are the largest contribution to the uncertainty in the jet energy. They are treated event by event and by an embedding procedure [2] while the final reconstructed transverse momenta are corrected by unfolding.

2. Dataset and unfolding

For this analysis, data collected in the heavy-ion runs of 2010 are used. For the jet reconstruction the anti-k_T algorithm from the FastJet package [3] with resolution parameters R=0.2 and R=0.3 is used. Since the electromagnetic calorimeter has only been fully installed in the beginning of 2011, jets use only charged tracks with $p_T > 0.150$ GeV and their 4-momentum vector is calculated using the boost invariant p_T recombination scheme. The Time Projection Chamber (TPC) and the Inner Tracking System (ITS) are the main detectors to measure the charged particle momenta in the pseudorapidity range $|\eta| <0.9$.

In the Pb-Pb collisions, the background from the Underlying Event (UE) contribution plays a very important role and must be subtracted from the reconstructed jet p_T. The summed transverse momentum from the UE is calculated as the product of the mean momentum density $\rho = median(p_T^{jet}/A^{jet})$ and the jet area A^{jet}.

The smearing introduced by the UE energy fluctuation can be evaluated by $\delta p_T = (p_T^{rec} - \rho A^{jet}) - p_T^{true}$ which give the difference between the UE corrected summed p_T and the true jet. The embedding method to determine δp_T is described in [2].

Moreover, background fluctuations have a large impact on the measured jet spectrum due to the tail to upwards fluctuations in the δp_T distribution. These fluctuations are corrected for by unfolding [4]. A response matrix $RM_{\delta p_T}$ containing the p_T smearing due to background fluctuations is constructed from the measured δp_T distribution.

Detector effects affecting the jet energy resolution have been studied with the detector simulation using PYTHIA and HIJING. The magnitude of these corrections results in a 10% shift in the jet energy scale which corresponds to 40% on the jet yield. The uncertainty on the exact knowledge of the tracking efficiency results in a 3% uncertainty on the jet energy scale reported in the systematic uncertainty of the measurement.

The two response matrices from background fluctuations and detector effects are combined to obtain the response matrix which will be used in the unfolding procedure: $M = RM_{\delta p_T} \cdot RM_{det} \cdot T$ in which M is the measured jet yield and T is the true jet yield.

Jet spectra are unfolded using a χ^2 minimization method. Using this method the number of jets is always conserved. The χ^2 function to be minimized indicates how well the unfolded distribution convoluted with the response matrix (the refolded spectrum) describes the measured spectrum. The χ^2 function used in this analysis is:

$$\chi^2 = \sum_{\text{refolded}} \left(\frac{y_{\text{refolded}} - y_{\text{measured}}}{\sigma_{\text{measured}}} \right)^2 + \beta \sum_{\text{unfolded}} \left(\frac{d^2 log\ y_{\text{unfolded}}}{d\ log\ p_T^2} \right)^2 . \tag{1}$$

The first summation term of equation 1 gives the χ^2 between the refolded spectrum and the measured jet spectrum. The second summation term of equation 1 is the penalty term which is used to regularize the unfolded solution and favors a local power law. Regularization is necessary to avoid heavily fluctuating solutions. The strength of the applied regularization β is tuned so as to make sure the regularization term is not dominant. In case the regularization is dominant the penalty term is equal to or larger than the χ^2 between the refolded and measured spectrum. In this case the refolded spectrum does not describe the measured spectrum. In case the regularization is too weak or too strong off-diagonal correlations in the Pearson coefficients extracted from the covariance matrix are observed.

The regularization adds a systematic uncertainty of $\sim$ 10% for central events and $\sim$ 4% for peripheral events to the unfolded yield.

3. Results

Fig.1 shows the charged jet yields as a function of p_T normalised by the number of collisions with a resolution parameter R=0.3. Comparing jet spectra for different centrality classes, a sizable suppression increasing with centrality can be observed.

In order to study the modification of the Pb-Pb spectra with respect to an incoherent superposition of binary nucleon-nucleon collision, the jet nuclear modification factor R_{AA}^{Pythia} is shown in Fig.2. As a reference, we use the jet spectrum from PYTHIA-Perugia0 [5] simulated at the same centre of mass energy. A strong jet suppression is observed for central events similar to the R_{AA} of inclusive charged particles.

These results show that the full jet energy is not contained in jets with radii R=0.2 or R=0.3 for the heavy-ion collisions. In Fig.3, the ratio R_{CP}, central to peripheral for R=0.3, shows the jet yield measured with respect to 50-80% centrality, used as a reference.

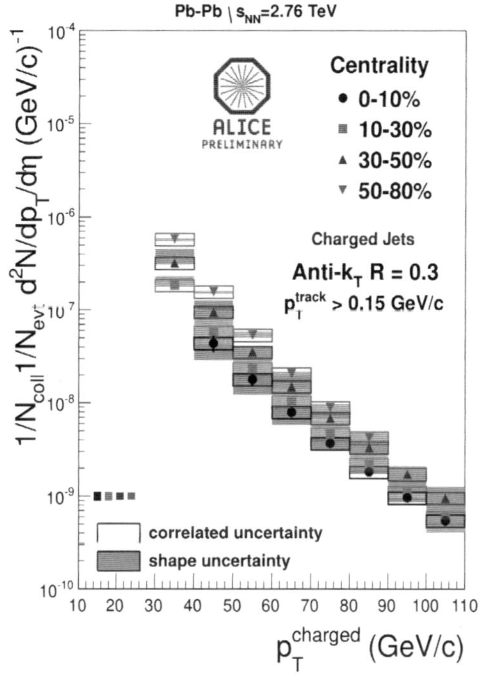

Figure 1. Unfolded jet spectra for jet radius $R = 0.3$.

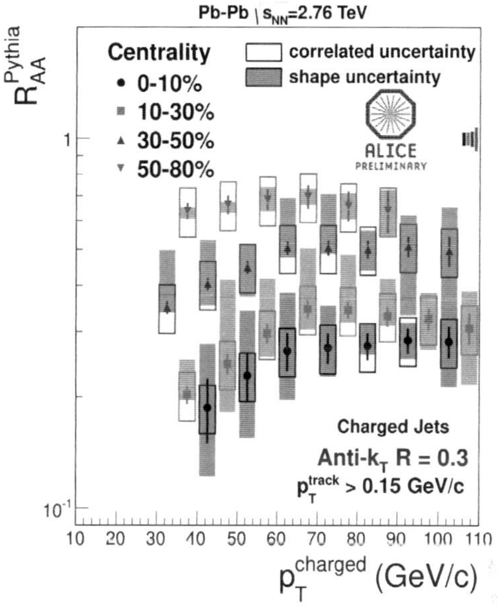

Figure 2. R_{AA}^{Pythia} for jet radius $R = 0.3$. For the reference , the jet spectrum from PYTHIA-Perugia0 is used.

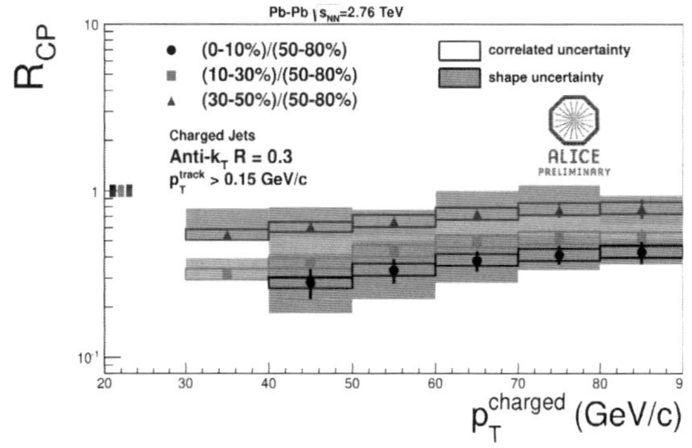

Figure 3. R_{CP} for jet radius $R = 0.3$. The peripheral jet spectrum corresponds to 50-80% centrality class.

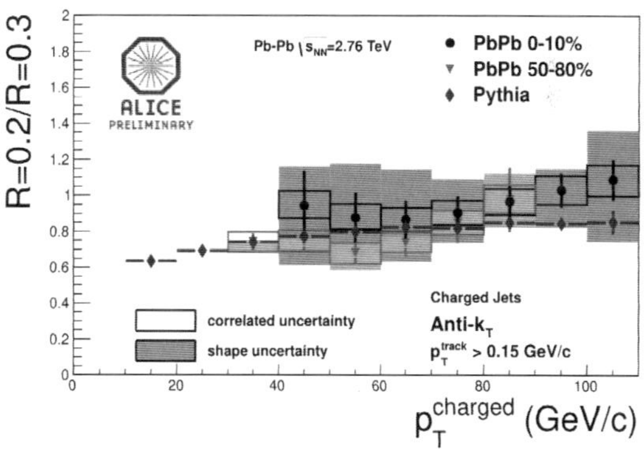

Figure 4. Ratio between $R = 0.2$ and $R = 0.3$ Pb-Pb jet spectra compared to PYTHIA.

Another important observable to measure is the ratio between the jet spectra for radii R=0.2 and R=0.3, as plotted in Fig.4. The result shows a jet production consistent with the one in vacuum for central to peripheral highlighting no significant jet broadening in the cross section ratio. A possible rise with p_T is shown due to the fact that higher p_T jets are more collimated. The data points are also in agreement with the pQCD and PYTHIA8 [6] calculation.

In Fig.5 and Fig.6, the charged jet results are compared to the JEWEL jet quenching MC

[7]. A good agreement is observed between the energy loss implementation of JEWEL and the charged jet results from ALICE.

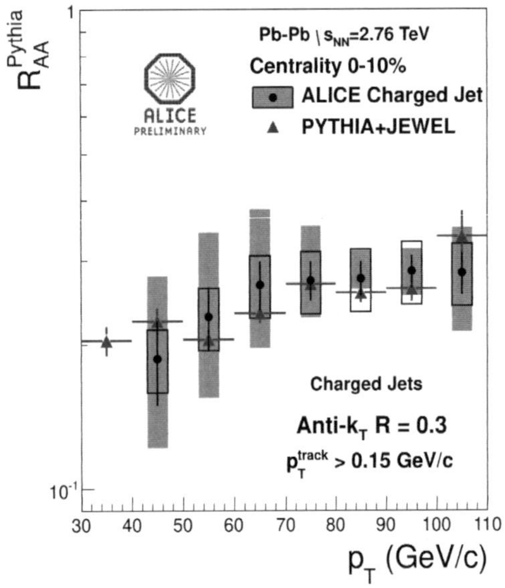

Figure 5. Comparison between JEWEL and measured charged jet R_{AA} for jet radius $R = 0.3$.

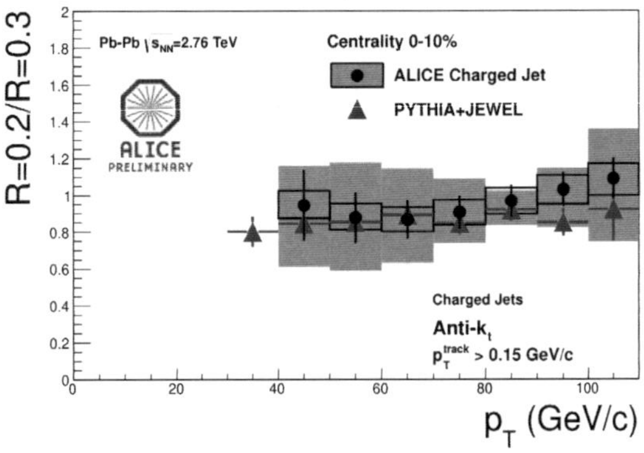

Figure 6. Ratio between $R = 0.2$ and $R = 0.3$ Pb-Pb jet spectra compared to JEWEL.

4. Conclusions

Important steps in the understanding of the heavy-ion collisions and the QGP physics have been achieved by ALICE. A strong suppression of the inclusive jet yield, in central collisions with respect to a PYTHIA *pp* reference, is observed. The measured suppression amounts to 0.20-0.35 in the momentum range 30-100 GeV. Measuring the ratio of jet yields at different resolutions, there is no indication of energy redistribution within the experimental uncertainties. These results are in qualitative agreement with models and are able to constrain or to exclude other models and calculations.

References

[1] C.Lourenco, H. Satz, Preface, Eur. Phys. J. C 43 (2005) 1; A.Majumder and M.Van Leeuwen, Prog.Part.Nucl.Phys. A66, 41 (2011);
[2] The ALICE Collaboration, B.Abelev et al., JHEP 1203 (2012) 053;
[3] M. Cacciari, G. P. Salam, G. Soyez, JHEP 0804 (2008) 063;
[4] M.Verweij for the ALICE Collaboration, arXiv:1208.6169;
[5] T. Sjostrand, S. Mrenna, P. Skands, PYTHIA 6.4 physics and manual, JHEP 05 (2006) 026 (arXiv:hep-ph/0603175);
[6] T. Sjostrand, S. Mrenna, P. Skands, Comput. Phys. Comm. 178 (2008) 852 (arXiv:0710.3820);
[7] K. C. Zapp, F. Krauss, U. A. Wiedemann, http://arxiv.org/abs/1111.6838.

International Conference on Heavy Ion Collisions in the LHC Era IOP Publishing
Journal of Physics: Conference Series **422** (2013) 012011 doi:10.1088/1742-6596/422/1/012011

Probing nuclear parton densities (and more) from $\gamma + Q$ production in p–A and A–A collisions

François Arleo
Laboratoire d'Annecy-le-Vieux de Physique Théorique (LAPTh), UMR5108, Université de Savoie, CNRS, BP 110, 74941 Annecy-le-Vieux cedex, France

Ingo Schienbein, Tzvetalina Stavreva
Laboratoire de Physique Subatomique et de Cosmologie, UJF, CNRS/IN2P3, INPG, 53 avenue des Martyrs, 38026 Grenoble, France

E-mail: arleo@lapp.in2p3.fr,schien@lpsc.in2p3.fr,stavreva@lpsc.in2p3.fr

Abstract. We present a detailed phenomenological study of the associated production of a prompt photon and a heavy-quark jet (charm or bottom) in proton-nucleus (p–A) collisions. The dominant contribution to the cross-section comes from the gluon–heavy-quark (gQ) initiated subprocess, making this process very sensitive to the gluon and the heavy quark nuclear parton densities. We show that the future p–A data to be collected at the LHC in this channel should help to disentangle the various nPDF sets currently available. In heavy-ion collisions, the photon transverse momentum can be used to gauge the initial energy of the massive parton propagating through the dense QCD medium produced in those collisions, making $\gamma + Q$ production a powerful process in order to probe energy loss dynamics in the heavy-quark sector.

1. $\gamma + Q$ as a versatile tool in nuclear collisions

The production of $\gamma + Q$ is a rich and versatile process in various hadronic collisions:

- In p–p and p–$\bar{\text{p}}$ collisions, first of all, $\gamma + Q$ production offers sensitive checks of perturbative QCD (pQCD) and might serve as a probe of intrinsic heavy-quark distributions inside the proton [1];
- In p–A collisions (e.g. at RHIC and soon at the LHC), this process can be used to constrain the gluon PDF in nuclei (nuclear PDF, nPDF), [2] which are pretty much unconstrained at small values of x. One should underline that knowing precisely nuclear PDF is a prerequisite in order to obtain reliable predictions in heavy-ion collisions;
- In A–A collisions, finally, the study of $\gamma + Q$ provides an ideal tool for investigating the energy lost by heavy quarks in the hot QCD medium produced in those collisions. Being an electromagnetic probe, the photon produced directly in the hard process is expected to traverse the medium unaffected. Its momentum can therefore serve as a proxy for the initial momentum of the heavy-quark propagating through the dense medium and eventually fragmenting into the heavy-quark jet. The imbalance between the prompt photon and the heavy-quark jet momentum from p–p to A–A collisions might thus reflect the amount of energy loss experienced by the heavy-quark. Furthermore, the comparison between $\gamma + c$ and $\gamma + b$ production would provide access to the mass hierarchy of parton energy loss.

In these proceedings we will discuss the constraints given by $\gamma+Q$ production in p–A collisions on nuclear parton densities. Energy loss effects on $\gamma+Q$ production in A–A collisions will also be briefly discussed.

2. Constraining the gluon nPDF through $\gamma + Q$ production

Unlike the PDF for a gluon inside a free proton, the nuclear gluon PDF is largely unconstrained due to the dearth of available data. Currently, only the NMC structure function data ($F_2^D(x, Q^2)$ and $F_2^{Sn}/F_2^C(x, Q^2)$) impose weak constraints on the gluon nPDF in the x-range $0.02 \lesssim x \lesssim 0.2$[1], so that a precise determination is not possible.

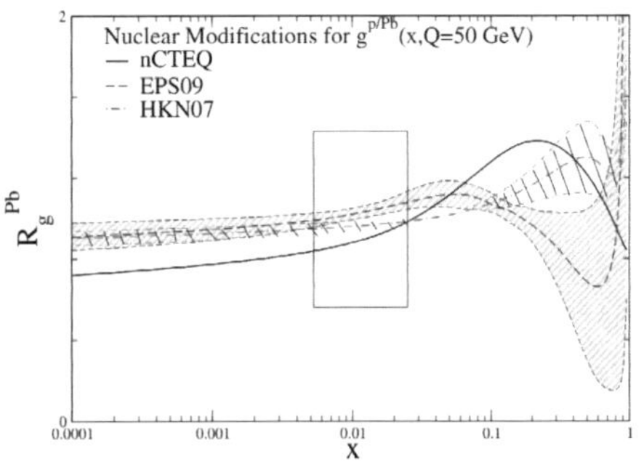

Figure 1. $R_g^{\rm Pb}(x, Q = x\sqrt{s}/2 \sim p_T)$ for nCTEQ [3], EPS09 [4] and HKN07 [5]. The box exemplifies the x-region probed at the LHC.

This large uncertainty in $g^{\rm A}(x, Q^2)$ is presented by the nuclear modification factor to the gluon nPDF, $R_g^{\rm Pb}(x, Q) = g^{\rm p/Pb}(x, Q)/g^{\rm p}(x, Q)$, in Figure 1 where a comparison between different nuclear PDF sets currently available (nCTEQ [3], HKN07 [5], EPS09 [4]) is shown.

The need for measurements of processes sensitive to the gluon nPDF is evident. Here we point out that $\gamma + Q$ production is an excellent probe of $g^A(x, Q^2)$, and can serve as one such process, as evidenced by Figure 2 and Figure 3. Figure 2 shows the differential cross-section for both γ+c and γ+b at $\sqrt{s_{\rm NN}} = 8.8$ TeV for p–Pb collisions in the ALICE EMCal acceptances. The anticipated event rate (before experimental efficiencies) is sufficiently large for a measurement ($N_{\gamma+{\rm c}}^{\rm pPb} = 1.2\times10^4$, $N_{\gamma+{\rm b}}^{\rm pPb} = 2.3\times10^3$). In Figure 3 the subprocess contributions to ${\rm d}\sigma_{\gamma+{\rm c}}^{\rm pPb}/{\rm d}p_{T\gamma}$ are presented, with gQ and gg fusion being the dominant ones; for more details see Ref. [2]. The sensitivity to the gluon nPDF further shows up in the nuclear modification factor to the cross-section, $R_{\rm pPb}^{\gamma+{\rm c}} = 1/208 \times {\rm d}\sigma/{\rm d}p_{T\gamma}({\rm pPb} \to \gamma + c + X)/{\rm d}\sigma/{\rm d}p_{T\gamma}({\rm pp} \to \gamma{+}{\rm c} + {\rm X})$ in Figure 4a, when compared to $R_g^{\rm Pb}(x, Q)$ in Figure 4b. It can clearly be seen by juxtaposing both figures that $R_{\rm pPb}^{\gamma+{\rm c}}$ follows closely $R_g^{\rm Pb}$ in the region of x probed at the LHC for each nPDF set. Therefore we can conclude that this process is an excellent candidate for constraining the gluon nuclear distribution as a measurement of the prompt photon + heavy-quark jet process with appropriately small error bars will be able to distinguish between the various nPDF sets.

3. Summary and outlook

Prompt photon + heavy-quark jet production is a versatile and potentially very useful process which proves highly sensitive to the heavy-quark and gluon content of the incoming hadrons.

[1] The EPS09 fit also includes data on π^0 production at RHIC.

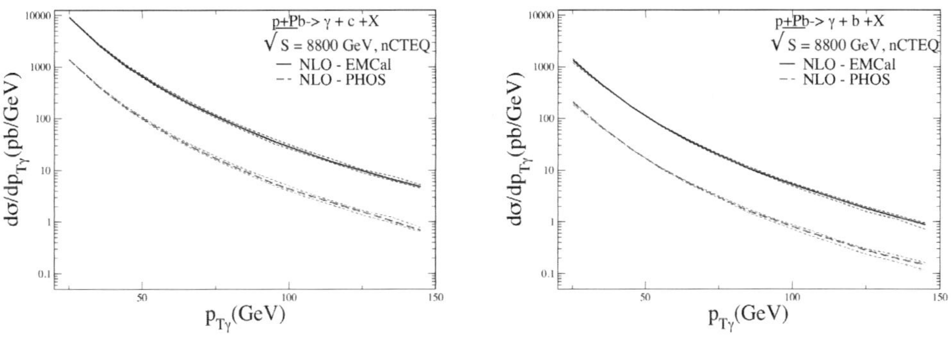

Figure 2. NLO differential cross-section for $\gamma + c$ (left) and $\gamma + b$ (right) production in p–Pb collisions at a center-of-mass energy of $\sqrt{s_{\rm NN}} = 8.8$ TeV in PHOS (lower band) and EMCal (upper band) acceptances.

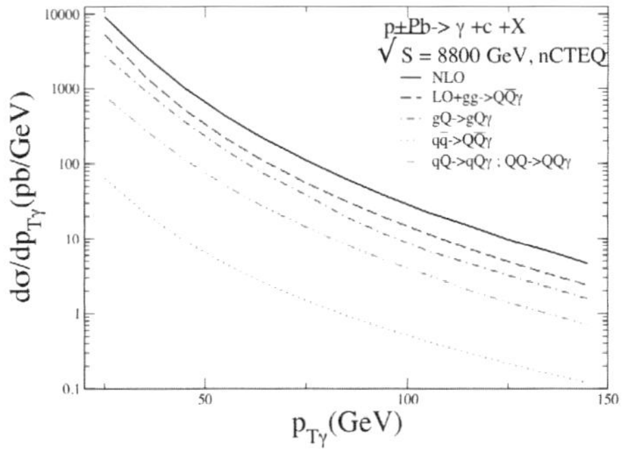

Figure 3. Subprocess contributions to the differential cross-section shown in Figure 2 (left), NLO (solid black line), LO+$gg \to Q\bar{Q}\gamma$ (dashed blue line), $gQ \to gQ\gamma$ (dash-dotted purple line), $q\bar{q} \to Q\bar{Q}\gamma$ (dotted red line), $qQ \to qQ\gamma$; $QQ \to QQ\gamma$ (dash-dot-dotted magenta line).

As a consequence, we showed in these proceedings that it can be used in p–A collisions in order to bring important constraints on heavy-quark and gluon nuclear PDF, both of which being presently poorly constrained.

In a recent work [6], we study the role of parton energy loss processes on the production of photon and heavy-quark jets in heavy-ion collisions. As an illustration, the quenching of the charm $p_\perp$ spectrum in $\gamma + c$ production has been computed as a function of p_{Tc} in Fig. 5a under various hypothesis regarding the energy loss of charm quarks, i.e. assuming the charm quarks to suffer the same energy loss as that of a light quark (dash-dotted line) or that of a heavy-quark assuming $m_Q = m_c = 1.5$ GeV (dashed) and $m_Q = m_b = 4.5$ GeV (solid) in the continuous part of the quenching weights (for completeness the calculation has also been carried out using the quenching weight of a propagating gluon, dotted line). The calculation is performed at leading order according to the Compton subprocess. The quenching factors computed using these various prescriptions follow closely the hierarchy of parton energy loss in the BDMPS-Z

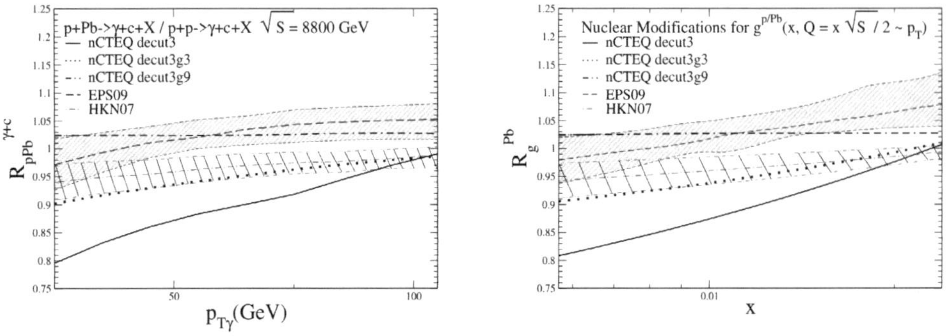

Figure 4. a) $R^{\gamma+c}_{\mathrm{pPb}}$ at LHC within ALICE EMCal acceptances, using nCTEQ [3], EPS09 [4] and HKN07 [5]. b) $R_g^{\mathrm{Pb}}(x, Q = x\sqrt{s}/2 \sim p_T)$ in the x region probed at the LHC.

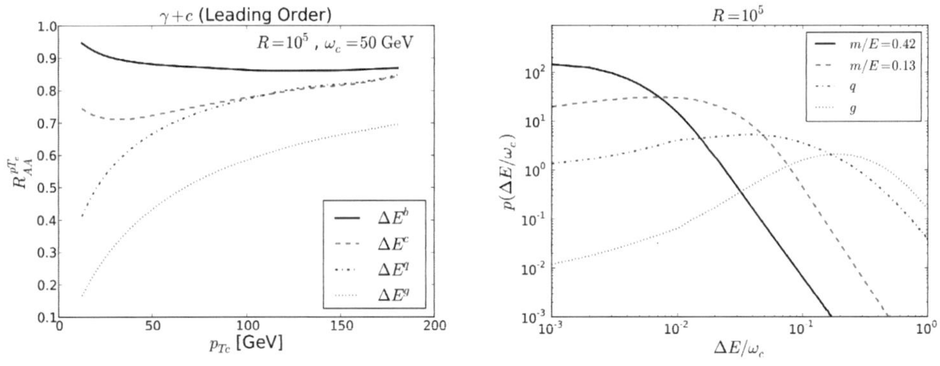

Figure 5. Left: Supression factors R_{AA} vs. p_{Tc}, applying energy loss according to the quenching weights shown in the right figure; Right: Quenching weights for energy loss corresponding to gluons (dotted, brown line), light quarks (dash-dotted, blue line),charm quarks (dashed, red line), and bottom quarks (solid, black line) in dependence of the energy loss scaled to ω_c.

quenching weights [7] used for each flavours in the calculation (shown in Fig. 5b). In particular, a much stronger (respectively, weaker) heavy-quark jet quenching would be expected should the charm quark lose energy like a gluon (respectively, bottom quark). Interesting is the comparison between the predictions assuming light quark and charm quark energy loss. At low transverse momenta, charm quark jet should be less suppressed than assuming light quark energy loss while at large momenta, $p_{Tc} \gg m_c$, the induced energy loss of a charm and a light quark should be identical and so does the quenching factor, as can be seen in Fig. 5a (compare the dotted and dash-dotted lines). This result and others are presented and discussed in [6].

[1] Stavreva T and Owens J 2009 *Phys. Rev.* **D79** 054017 (*Preprint* 0901.3791)
[2] Stavreva T *et al.* 2011 *JHEP* **01** 152 (*Preprint* 1012.1178)
[3] Schienbein I *et al.* 2009 *Phys. Rev.* **D80** 094004 (*Preprint* 0907.2357)
[4] Eskola K J, Paukkunen H and Salgado C A 2009 *JHEP* **04** 065 (*Preprint* 0902.4154)
[5] Hirai M, Kumano S and Nagai T H 2007 *Phys. Rev.* **C76** 065207 (*Preprint* 0709.3038)
[6] Stavreva T, Arleo F and Schienbein I (*Preprint* 1211.6744)
[7] Armesto N, Salgado C A and Wiedemann U A 2004 *Phys. Rev.* **D69** 114003 (*Preprint* hep-ph/0312106)

Journal of Physics: Conference Series **422** (2013) 012012 doi:10.1088/1742-6596/422/1/012012

J/Ψ in nuclear matter

G. Krein

Instituto de Física Teórica, Universidade Estadual Paulista
Rua Dr. Bento Teobaldo Ferraz, 271 - Bloco II, 01140-070 São Paulo, SP, Brazil

E-mail: gkrein@ift.unesp.br

Abstract. We report on recent estimates of the J/Ψ mass shift in infinite nuclear matter and finite nuclei arising from in-medium D and D^* meson loops. The density dependence of the J/Ψ mass shift is evaluated employing medium-modified D and D^* meson masses derived within the quark-meson coupling model. Using a local density approximation, J/Ψ-nuclear bound state energies are calculated for a range of nuclei. We predict that J/Ψ-nuclear bound states should be observed with a clear signal in experiments, provided the J/Ψ meson is produced in recoilless kinematics.

1. Introduction

New opportunities will be opened for studying properties of charmonia and charmed mesons in the nuclear medium with the 12 GeV upgrade of the CEBAF accelerator at the Jefferson Lab in the USA and with the construction of the FAIR facility in Germany. These new facilities will have the potential of implanting low-momentum charmed mesons like J/Ψ, η_c, D and D^*. Particularly exciting perspectives are the possibility of creating new exotic nuclear bound states by nuclei capturing charmonia states like J/Ψ and η_c [1, 2, 3, 4], or heavy-light D and D^* mesons [5, 6, 7]. Capture of J/Ψ (and other charmonia) by nuclei is particularly interesting because a charmonium state does not have quarks in common with the nuclear medium and therefore its interactions with the medium necessarily involve the intervention of gluons. Basic interaction mechanisms discussed in the literature have been the excitation of QCD van der Waals forces arising from the exchange of two or more gluons between color-singlet states, and the excitation of charmed hadronic intermediate states with light quarks created from the vacuum.

In the present communication we present recent results for the mass shift of J/Ψ in terms of the excitation of intermediate charmed mesons using effective Lagrangians. In addition to the $D\bar{D}$ loops, we also include $D\bar{D}^*$, $D^*\bar{D}$ and $D^*\bar{D}^*$ loops. The medium dependence of the D and D^* masses in nuclear matter is obtained by an explicit calculation using the quark-meson coupling (QMC) model – for a review of the model see Ref. [8]. The QMC is a quark-based model for nuclear structure which has been very successful in describing nuclear matter saturation properties and has been used to predict a great variety of changes of hadron properties in nuclear medium. We also present predictions for J/Ψ energy levels in selected nuclei, obtained by solving a Klein-Gordon for the J/Ψ wave function in presence of an effective local potential derived in a local density approximation from the mass shift in infinite nuclear matter.

2. The model

Initially we specify the phenomenological Lagrangian densities for the vertices J/Ψ-D and J/Ψ-D^* (ψ denotes the field representing J/Ψ):

$$\mathcal{L}_{\psi DD} = ig_{\psi DD}\, \psi^\mu \left[\bar{D}\,(\partial_\mu D) - (\partial_\mu \bar{D})\, D\right], \quad (1)$$

$$\mathcal{L}_{\psi DD^*} = \frac{g_{\psi DD^*}}{m_\psi}\, \varepsilon_{\alpha\beta\mu\nu} \left(\partial^\alpha \psi^\beta\right) \left[\left(\partial^\mu \bar{D}^{*\nu}\right) D + \bar{D}\left(\partial^\mu D^{*\nu}\right)\right], \quad (2)$$

$$\mathcal{L}_{\psi D^* D^*} = ig_{\psi D^* D^*} \left\{\psi^\mu \left[\left(\partial_\mu \bar{D}^{*\nu}\right) D^*_\nu - \bar{D}^{*\nu}\left(\partial_\mu D^*_\nu\right)\right] + \left[\left(\partial_\mu \psi^\nu\right) \bar{D}^*_\nu - \psi^\nu \left(\partial_\mu \bar{D}^*_\nu\right)\right] D^{*\mu} + \bar{D}^{*\mu}\left[\psi^\nu \left(\partial_\mu D^*_\nu\right) - \left(\partial_\mu \psi^\nu\right) D^*_\nu\right]\right\}. \quad (3)$$

These Lagrangians are an $SU(4)$ extension of light-flavor chiral-symmetric Lagrangians of pseudoscalar and vector mesons. Note that we are not using a local gauge symmetry principle for vector mesons [2] which would introduce contact interactions involving two pseudoscalar and two vector mesons – this would introduce an additional contact interaction of the form $2g^2_{\psi DD}\psi^\mu\psi_\mu \bar{D}D$ in Eq. (1), leading to a small positive (instead of a small negative) mass shift due to the DD loop [3]. Given the Lagrangians, the mass-difference $\Delta m = m^*_\psi - m_\psi$, where m_ψ and m^*_ψ are respectively the vacuum and in-medium masses of J/Ψ, defined as:

$$m^2_\psi = (m^0_\psi)^2 + \Sigma(k^2 = m^2_\psi) \qquad \text{and} \qquad m^{*2}_\psi = (m^0_\psi)^2 + \Sigma^*(k^2 = m^{*2}_\psi), \quad (4)$$

where m^0_ψ is the bare mass and $\Sigma(k^2)$ is the total J/Ψ self-energy obtained by summing the contributions from the DD, DD^* and D^*D^* loops (we take $m_{D^0} = m_{D^\pm}$ and $m_{D^{*0}} = m_{D^{*\pm}}$):

$$\Sigma_l(m^2_\psi) = -\frac{g^2_{\psi l}}{3\pi^2} \int_0^\infty dq\, \mathbf{q}^2\, F_l(\mathbf{q}^2)\, K_l(\mathbf{q}^2), \quad (5)$$

where $F_l(\mathbf{q}^2)$ is the product of vertex form-factors and the $K_l(\mathbf{q})$ for each loop contribution are given by

$$K_{DD}(\mathbf{q}^2) = \frac{\mathbf{q}^2}{\omega_D}\frac{\mathbf{q}^2}{\omega^2_D - m^2_\psi/4}, \qquad K_{DD^*}(\mathbf{q}^2) = \frac{\mathbf{q}^2\,\overline{\omega}_D}{\omega_D\omega_{D^*}}\frac{1}{\overline{\omega}^2_D - m^2_\psi/4}, \quad (6)$$

$$K_{D^*D^*}(\mathbf{q}^2) = \frac{1}{4m_\psi\omega_D}\left[\frac{\mathcal{A}(q^0 = \omega_{D^*})}{\omega_{D^*} - m_\psi/2} - \frac{\mathcal{A}(q^0 = \omega_{D^*} + m_\psi)}{\omega_{D^*} + m_\psi/2}\right], \quad (7)$$

where $\omega_D = (\mathbf{q}^2 + m^2_D)^{1/2}$, $\omega_{D^*} = (\mathbf{q}^2 + m^2_{D^*})^{1/2}$, $\overline{\omega}_D = (\omega_D + \omega_{D^*})/2$, $\xi = 0$ for the non-gauged Lagrangian of Eq. (1) and $\xi = 1$ for the gauged Lagrangian of Ref. [2], and $\mathcal{A}(q) = \sum_{i=1}^4 \mathcal{A}_i(q)$, with

$$\mathcal{A}_1(q) = -4q^2\left\{4 - \frac{q^2 + (q-k)^2}{m^2_{D^*}} + \frac{[q\cdot(q-k)]^2}{m^4_{D^*}}\right\}, \quad (8)$$

$$\mathcal{A}_2(q) = 8\left[q^2 - \frac{q\cdot(q-k)}{m^2_{D^*}}\right]\left[2 + \frac{(q^0)^2}{m^2_{D^*}}\right], \quad (9)$$

$$\mathcal{A}_3(q) = 8\left(2q^0 - m_\psi\right)\left\{q^0 - \left(2q^0 - m_\psi\right)\frac{q^2 + q\cdot(q-k)}{m^2_{D^*}} + q^0\frac{[q\cdot(q-k)]^2}{m^4_{D^*}}\right\}, \quad (10)$$

$$\mathcal{A}_4(q) = -8\left[q^0 - (q^0 - m_\psi)\frac{q\cdot(q-k)}{m^2_{D^*}}\right]\left[(q^0 - m_\psi) - q^0\frac{q\cdot(q-k)}{m^2_{D^*}}\right]. \quad (11)$$

Here, q and k are four-vectors given by $q = (q^0, \mathbf{q})$ and $k = (m_\psi, 0)$. The expression for Σ^* is identical to Σ, but calculated with in-medium D and D^* masses.

The medium dependence of the masses of the D and D^* mesons are obtained using the QMC model. Explicitly, the masses are given by

$$m^*_{D,D^*} = \sum_{j=q,\bar{q},c,\bar{c}} \frac{n_j \Omega^*_j - z_{D,D^*}}{R^*_{D,D^*}} + \frac{4}{3}\pi B R^{*3}_{D,D^*}, \tag{12}$$

where B is the bag constant, $\Omega^*_q = \Omega^*_{\bar{q}} = [x_q^2 + (R^*_{D,D^*} m^*_q)^2]^{1/2}$ $(q = u, d)$, with $m^*_q = m_q - g^q_\sigma \sigma$, $\Omega^*_c = \Omega^*_{\bar{c}} = [x_c^2 + (R^*_{D,D^*} m_c)^2]^{1/2}$, $x_{q,c}$ are the bag eigenfrequencies and $n_q(n_{\bar{q}})$ and $n_c(n_{\bar{c}})$ are the lowest mode quark (antiquark) numbers for the quark flavors q and c in the D and D^* mesons, respectively, and the z_{D,D^*} parameterize the sum of the center-of-mass and gluon fluctuation effects and are assumed to be independent of density. The bag radii R^*_D and $R^*_{D^*}$ are obtained from the minimization of the masses, $\partial m^*_{D,D^*}/\partial R^*_{D,D^*} = 0$.

Using a local density approximation, one can identify $m^*_\psi(\rho_B) - m_\psi \equiv U_\psi(r)$, with $\rho_B = \rho_B(r)$ being the nuclear density, as an effective potential for the J/Ψ in a nucleus. Energy levels are then obtained solving a Klein-Gordon equation for the J/Ψ wave function $\phi_\psi(\vec{r})$:

$$\left[\nabla^2 + E_\psi^2 - \mu^2 - 2\mu U_\psi(r)\right] \phi_\psi(\vec{r}) = 0, \tag{13}$$

where E_Ψ is the total energy of the J/Ψ meson, and $\mu = m_\Psi m_A/(m_\Psi + M)$ with m_Ψ (M) being the vacuum mass of the J/Ψ meson (nucleus). Further details can be found in Ref. [4].

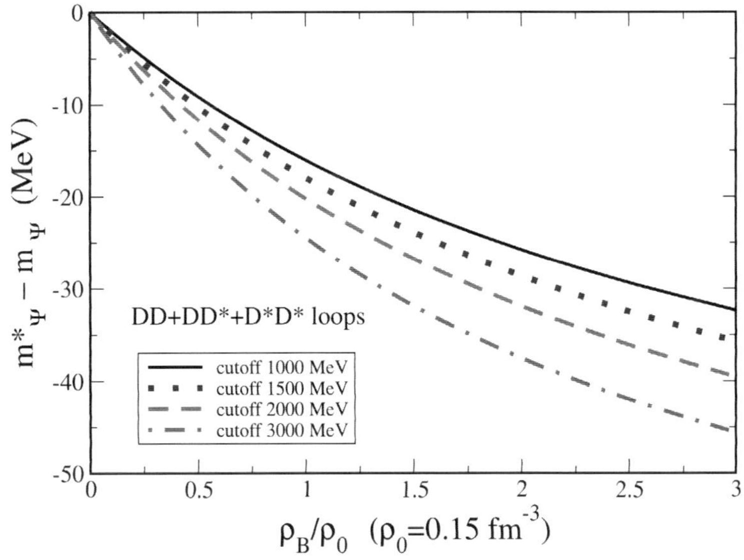

Figure 1. J/Ψ mass difference as a function of nuclear matter density. Figure from Ref. [3].

3. Results

We choose the values $(m_q, m_c) = (5, 1300)$ MeV for the current quark masses, $B = 170$ MeV^4 and $R_N = 0.8$ fm for the bag radius of the nucleon in free space. The quark-meson coupling constants, g^q_σ, g^q_ω and g^q_ρ, are adjusted to fit the nuclear saturation energy and density of symmetric nuclear matter, and the bulk symmetry energy [8]. Exactly the same coupling constants, g^q_σ, g^q_ω and g^q_ρ, are used for the light quarks in the D and D^* mesons and baryons as in the nucleon. The nuclear densities $\rho_B(r)$ are calculated within the QMC with the same set of parameters. Another ingredient of the calculation are form factors, necessary to regularize the loop integrals. We parametrize them as dipoles with a cutoff mass Λ - we use the same Λ in all vertices. In our estimates we vary Λ over a wide range of values to test sensitivity of results to this crucial input.

In Fig. 1 we present results for the mass shift as a function of the nuclear matter density. Clearly seen, is a negative shift in the mass of J/Ψ in nuclear matter, of the order of $10-20$ MeV. As estimated in Ref. [3], such an attraction is more than enough to bind a J/Ψ meson to a nucleus. Indeed, solving the Klein-Gordon equation of Eq. (13) for a series of nucleus, ranging from small nuclei like ^{4}He and ^{12}C to large ones like ^{90}Zr and ^{208}Pb, one finds [4] single-particle energy levels with sizable binding energies, as shown in Table 1 for selected nuclei. Such J/Ψ-nuclear bound states should be observed experimentally, provided, of course, the J/Ψ meson is produced in recoilless kinematics. The model leaves room for improvements, like a systematic study of the interaction of the D mesons with nucleons [9, 10].

Table 1. Energy levels of J/Ψ in selected nuclei - results from Ref. [4].

		$\Lambda_{D,D^*} = 1500$ MeV	$\Lambda_{D,D^*} = 2000$ MeV
		E (MeV)	E (MeV)
$^4_\Psi$He	1s	-4.19	-5.74
$^{40}_\Psi$Ca	1s	-14.96	-17.24
	1p	-10.81	-12.92
	1d	-6.29	-8.21
	2s	-5.63	-7.48
$^{208}_\Psi$Pb	1s	-16.83	-19.10
	1p	-15.36	-17.59
	1d	-13.61	-15.81
	2s	-13.07	-15.26

Acknowledgments
I would like to thank Ding Lu, Tony Thomas and Kazuo Tsushima for the collaborative work. Work partially financed by CNPq and FAPESP (Brazilian agencies).

References

[1] Brodsky S J, Schmidt I A and de Teramond G 1990 *Phys. Rev. Lett.* **64** 1011
[2] Lee S H and Ko C M 2003 *Phys. Rev.* C **67** 038202
[3] Krein G, Thomas A W and Tsushima K 2011 *Phys. Lett.* B **697** 136
[4] Tsushima K, Lu D H, Krein G and Thomas A W 2011 *Phys. Rev.* C **83** 065208
[5] Tsushima K, Lu D H, Thomas A W, Saito K and Landau R H 1999 *Phys. Rev.* C **59** 2824
[6] Garcia-Recio C, Nieves J and Tolos L 2010 *Phys. Lett.* B **690** 369
[7] Garcia-Recio C, Nieves L, Salcedo L L and Tolos L 2012 *Phys. Rev.* C **85** 025203
[8] Saito K, Tsushima K and Thomas A W 2007 *Prog. Part. Nucl. Phys.* **58** 1
[9] Haidenbauer J, Krein J, Meissner U G and Sibirtsev A 2008 *Eur. Phys. J.* A **37** 55
[10] Haidenbauer J, Krein G, Meissner U G and Tolos L 2011 *Eur. Phys. J.* A **47** 18

Journal of Physics: Conference Series **422** (2013) 012013 doi:10.1088/1742-6596/422/1/012013

Heavy Flavor Results from STAR

Qiu Hao (for the STAR Collaboration)
Lawrence Berkeley National Lab, 1 Cyclotron Road, Berkeley, CA 94720, U.S.A.
E-mail: hqiu@lbl.gov

Abstract. Heavy flavor quarks are considered to be unique probes of the medium created in high energy heavy ion collisions. These proceedings report selected STAR results from various measurements of heavy flavor production, including J/ψ suppression and collectivity, Υ suppression, ratio of bottom-decay electrons to charm-decay electrons, D meson spectra and charm cross section.

1. Introduction

The mass of heavy flavor quarks is significantly higher than the critical temperature at RHIC energies, Λ_{QCD} or the mass of u, d, and s quarks. As a consequence, heavy quarks are produced by hard processes early in the collision. Because heavy quarks decay weakly to light quarks within a much longer time than the collision, they can be treated as conserved in total number. Additionally, they are less influenced by the later hadronic stage due to their large mass. These features make heavy quarks an ideal probe for studying the QGP medium properties at RHIC.

In the following, the J/ψ nuclear modification factor R_{AA} and elliptic flow v_2, the Υ nuclear modification factor, the ratio of bottom-decay electrons to charm-decay electrons in p+p collisions, p_T spectra of D^0 and D^* mesons, and the charm cross section deduced from them will be presented.

2. J/ψ nuclear modification factor and elliptic flow

In relativistic heavy-ion collisions, the $c\bar{c}$ bound state is subject to dissociation due to the color screening of the binding potential in the deconfined medium. As a consequence, the J/ψ production is expected to be suppressed, and such suppression has been proposed as a signature of QGP formation [1]. However, the J/ψ yield in heavy ion collisions can also be affected by other hot and cold nuclear effects [2, 3, 4, 5]. Precise measurements of J/ψ R_{AA} as well as v_2 as a function of p_T and centrality will help to understand J/ψ production and whether J/ψ is suppressed by the color screening effect.

Figure 1 shows J/ψ R_{AA} vs. N_{part} for different sets of data and for different model calculations at $\sqrt{s_{NN}} = 200$ GeV [6, 7]. We observe more suppression in central than in peripheral collisions, and R_{AA} is larger for $p_T > 5$ GeV/c than without p_T cut. In peripheral collisions, R_{AA} is consistent with unity at high p_T, while in central collisions, the results show a significant suppression even in the high p_T region. The lines in the plot represent predictions from two models that consider J/ψ from both initial pQCD production and coalescence production in transport model calculations [8, 9]. The green lines are for the case where no p_T cut is applied, while the blue lines are for high p_T ranges. The models, which include color screening effects, can describe the general trend of the data points.

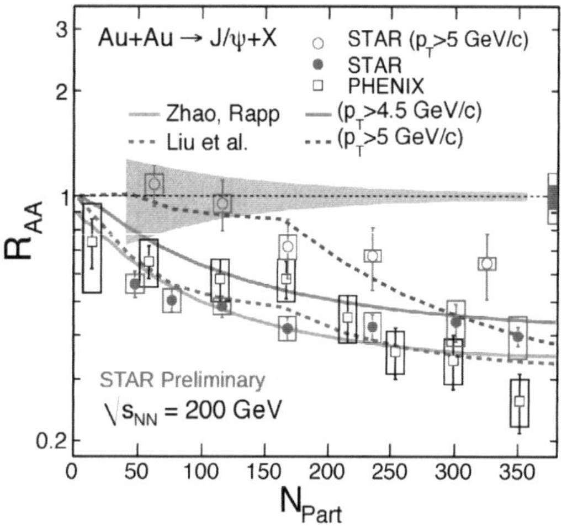

Figure 1. J/ψ R_{AA} as a function of N_{part}.

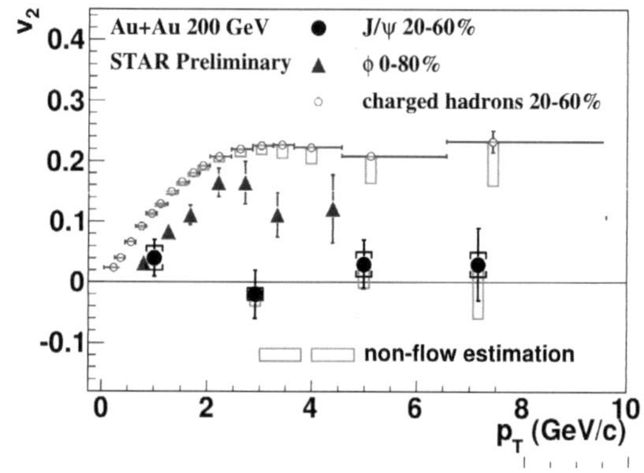

Figure 2. v_2 as a function of p_T for J/ψ, ϕ and charged hadrons.

Elliptic flow (v_2) constrains the J/ψ production mechanism from another dimension. J/ψ produced from direct pQCD processes should have little azimuthal preference, while J/ψ produced from recombination of thermalized charm quarks will inherit the flow of charm quarks, exhibiting sizable flow [10]. Flow of J/ψ from recombination can also be used to probe the charm quark flow, which provides a unique tool to test thermalization of the medium.

In Figure 2 J/ψ v_2 for the 20 - 60 % most central Au+Au collisions at 200 GeV is presented as a function of p_T [11]. For reference, v_2 of charged hadrons and of ϕ mesons is also shown. The J/ψ v_2 at $p_T > 2$ GeV/c is consistent with zero within errors, in contrast to the results for light quark hadrons. This result disfavors the scenario that J/ψ at large p_T are produced dominantly by coalescence from thermalized charm and anti-charm quarks, and is consistent with predictions from models in Figure 1, which consider both initial and coalescence production.

3. Υ nuclear suppression factor

Compared to J/ψ, Υ states are much cleaner probes to test suppression due to the screening effect because recombination can be neglected at RHIC [12] and there are negligible weak decay feed-down. Figure 3 shows Υ (1S+2S+3S) R_{AA} as a function of N_{part}. We observe an increase of suppression with larger N_{part}. And the measured R_{AA} in the most central collisions is consistent with the prediction from the internal energy potential model requiring strong 2S and complete 3S suppression [13].

4. Non-photonic Electrons (NPE) from bottom and charm

Open heavy flavor particles are also important in studying the medium created in heavy ion collisions. Open charm can be used to measure the total charm yield, which is an important base line for understanding charmonia coalescence production. Furthermore, the R_{CP} and R_{AA} of charm and bottom as a function of p_T can be used to study heavy quark energy loss in QGP. Open heavy flavor particles decay before reaching the detectors. One way to study them is to measure their decay electrons. Heavy flavor, photon conversion and Dalitz decays are the major sources of electron production at RHIC. By reconstructing and subtracting the latter two sources, STAR can measure NPE from bottom and charm decays [14, 15].

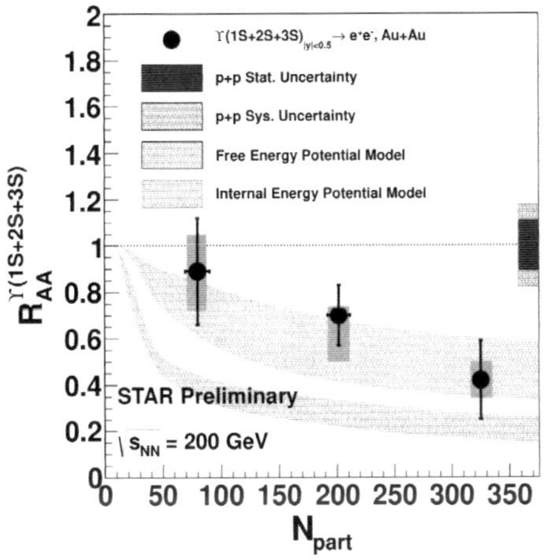

Figure 3. Υ R_{AA} vs. N_{part}.

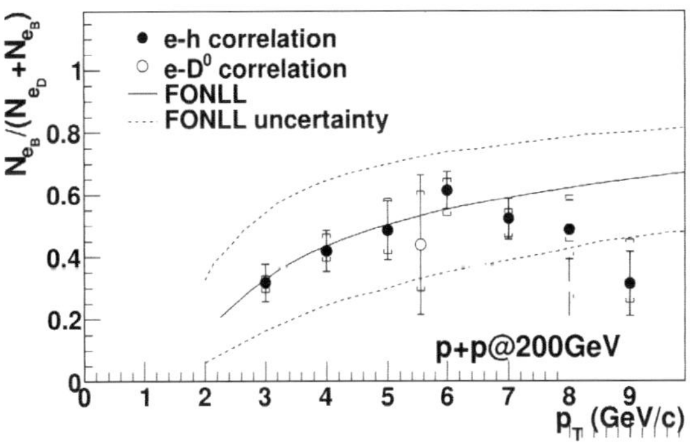

Figure 4. Ratio of NPE from B to all NPE.

The ratio of bottom-decay electrons to total NPE in p+p 200 GeV collisions can be obtained by fitting the NPE-hadron azimuthal correlation data with correlation line shapes of charm and bottom decay electrons from PYTHIA. The result as a function of p_T is shown in Figure 4 [16]. We can see that the contribution of B meson decay increases with p_T. The result is consistent with FONLL prediction. A similar measurement of NPE-D^0 correlation gets consistent result.

5. D^0 and D^* p_T spectra and charm cross section

Measuring only the decay electrons does not constrain the heavy flavor kinematics due to momentum smearing. STAR also measured fully reconstructed D^0 and D^* mesons directly. Since the D mesons have a very short decay length, STAR can not distinguish their decay vertices from the primary vertex right now. A statistics cummulant method is used instead to beat the small signal to background (s/b) ratio. Figure 5 shows D^0 and D^* p_T spectra in p+p 200 GeV collisions divided by their predicted portion of total charm yield, 0.56 and 0.22 respectively [17, 18]. The result is consistent with the FONLL [19] upper limit.

From p_T spectra of D mesons we can deduce the charm cross section. This is also done for d+Au and Au+Au collisions with different centralities. Figure 6 shows the charm cross section per nucleon-nucleon collision as a function of number of binary collisions (N_{bin}) [18, 20]. We can see that the charm cross section follows N_{bin} scaling, which confirms that charm quarks are produced mostly via initial hard scatterings.

6. Conclusion and outlook

In summary, by using various heavy flavor probes, we gain a better understanding of the medium created in relativistic heavy ion collisions. J/ψ suppression pattern and elliptic flow results are consistent with models with both suppressed initial J/ψ and coalescence production in transport model. Υ R_{AA} shows more suppression in central collisions, and is consistent with a model requiring strong 2S and complete 3S suppression in central collisions. By measuring the azimuthal correlation between NPE and hadrons, an increasing ratio of NPE from bottom to total NPE as a function of p_T is observed in p+p collisions. The charm cross section measured through D mesons follows number of binary collisions scaling, which confirms that charm quarks are produced mostly via initial hard processes.

In the near future, STAR will have two major detector updates targeting precise heavy flavor

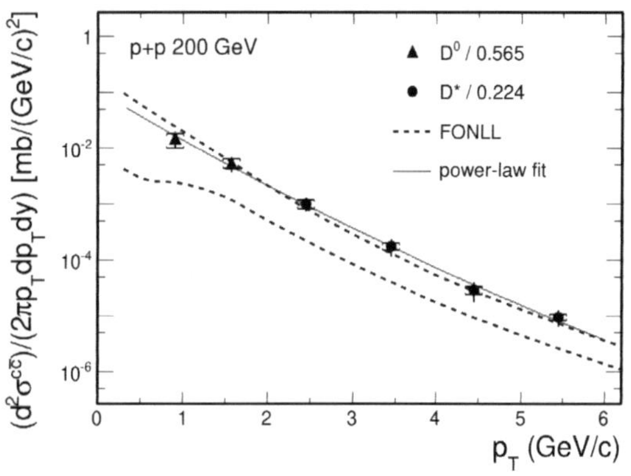

Figure 5. Scaled D^0 and D^* p_T spectra.

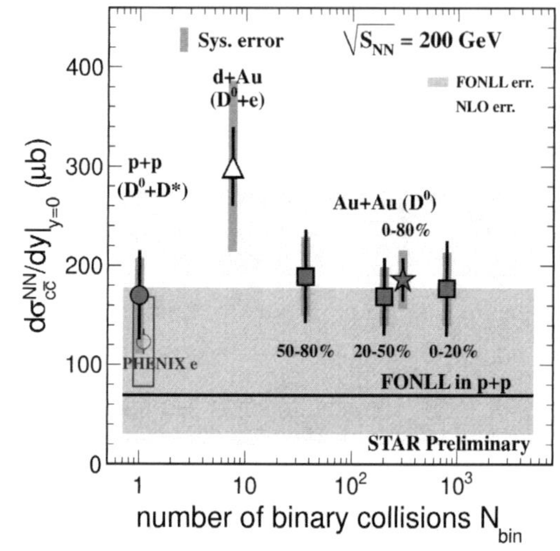

Figure 6. Charm cross section vs. N_{bin}.

measurements: the Heavy Flavor Tracker (HFT) and the Muon Telescope Detector (MTD). The HFT is a silicon vertex detector which will greatly reduce background for open heavy flavor particles by reconstructing secondary vertices. It will significantly enhance the s/b ratio of D and Λ_C measurements, and will be able to distinguish between prompt and B feed down J/ψ. It will also enable the measurement of the ratio of NPE from bottom to NPE from charm in Au+Au collisions. The MTD uses the magnet yoke as an absorber for other charged particles and detects muons, which can enhance J/ψ and Υ measurements with its fast trigger ability, and distinguish different Υ states in the di-muon decay channel with good mass resolution.

References

[1] T.Matsui, H. Satz, Phys. Lett. B 178, 416 (1986).
[2] R.L. Thews, Eur. Phys. J. C43, 97 (2005).
[3] A. Andronic et al., Nucl. Phys. A 789, 334 (2007).
[4] A. Capella et al., Eur. Phys. J. C58, 437 (2008).
[5] A. Adare et al. (PHENIX Collaboration), Phys. Rev. Lett. 107, 142301 (2011).
[6] L. Adamczyk et al. (STAR Collaboration), arXiv:1208.2736 (2012).
[7] C. Powell (for the STAR Collaboration), EPS HEP Conference, July 21-27, 2011.
[8] Y. Liu et al., Phys. Lett. B 678, 72 (2009) and private communication.
[9] X. Zhao, R. Rapp, Phys. Rev. C 82, 064905 (2010) and private communication.
[10] L. Yan, P. Zhuang, N. Xu, Phys. Rev. Lett. 97, 232301 (2006).
[11] H. Qiu (for the STAR Collaboration), Acta Physica Polonica B, Proceedings Supplement Vol. 5, No. 2, 323 (2012).
[12] Grandchamp, Sun, Van Hess, Rapp, Phys. Rev. C 73, 064906 (2006).
[13] Strickland and D. Bazow, artXiv:1112.2761v4.
[14] B. I. Abelev et al. (STAR Collaboration), Phys. Rev. Lett. 98, 192301 (2007).
[15] B. I. Abelev et al. (STAR Collaboration), Phys. Rev. Lett. 106, 159902(E) (2011).
[16] M. M. Aggarwal et al. (STAR Collaboration), Phys. Rev. Lett. 105, 202301 (2010).
[17] C. Amsler et al. (PDG), Phys. Lett. B 667.1 (2008).
[18] L. Adamczyk et al. (STAR Collaboration), arXiv:1204.4244.
[19] M. Cacciari, Phys. Rev. Lett. 95, 122001 (2005).
[20] Y. Zhang (for the STAR Collaboration), J. Phys. G 38, 124142 (2011).

Journal of Physics: Conference Series **422** (2013) 012014 doi:10.1088/1742-6596/422/1/012014

Heavy-flavour and quarkonium measurements in Pb–Pb collisions at $\sqrt{s_{\rm NN}} = 2.76$ TeV with ALICE

N. Bastid for the ALICE Collaboration

Laboratoire de Physique Corpusculaire, Clermont Université, Université Blaise Pascal, CNRS-IN2P3, Clermont-Ferrand, France

E-mail: nicole.bastid@clermont.in2p3.fr

Abstract. We report on the latest results on heavy-flavour and J/ψ production at both mid-rapidity and forward rapidity, in Pb–Pb collisions at $\sqrt{s_{\rm NN}} = 2.76$ TeV, measured with the ALICE experiment at the LHC. We present measurements of the nuclear modification factor for open heavy-flavours and J/ψ and compare these results to model predictions. Preliminary results on the elliptic flow of D mesons and J/ψ are discussed.

1. Introduction

ALICE [1] is the dedicated heavy-ion experiment at the LHC, optimized to study the properties of strongly interacting matter in the extreme conditions of high temperature and energy density expected to be reached. Heavy quarks (charm and beauty), abundantly produced at the LHC, are regarded as effective probes of this medium. Due to their large masses, they are created mainly in hard scattering processes in the early stage of the collision and subsequently interact with this medium. In particular, open heavy-flavour hadrons are expected to be sensitive to the energy density of the system through the mechanism of in-medium energy loss of heavy quarks, while quarkonia should provide an estimate of the medium temperature through their dissociation due to color screening [2]. The in-medium effects are usually quantified by means of the nuclear modification factor $R_{\rm AA}$ defined as:

$$R_{\rm AA}(p_{\rm T}) = \frac{1}{\langle T_{\rm AA}\rangle} \times \frac{{\rm d}N_{\rm AA}/{\rm d}p_{\rm T}}{{\rm d}\sigma_{\rm pp}/{\rm d}p_{\rm T}}, \qquad (1)$$

where $\langle T_{\rm AA}\rangle$ is the average nuclear overlap function in a given centrality class. ${\rm d}N_{\rm AA}/{\rm d}p_{\rm T}$ is the $p_{\rm T}$-differential yield in nucleus-nucleus (AA) collisions, while ${\rm d}\sigma_{\rm pp}/{\rm d}p_{\rm T}$ is the $p_{\rm T}$-differential cross section in pp collisions. According to QCD, quarks should lose less energy than gluons. In addition, heavy quarks are expected to lose less energy than light quarks due to the dead-cone effect [3]. Furthermore, the J/ψ suppression could be counteracted by regeneration processes which may become important at the LHC. The measurement of the elliptic flow v_2 of heavy-flavour particles and J/ψ should carry additional information on the medium transport properties. It is expected to provide insights on the possible degree of thermalization of heavy quarks in the medium at low $p_{\rm T}$ and on the path length dependence of energy loss at high $p_{\rm T}$.

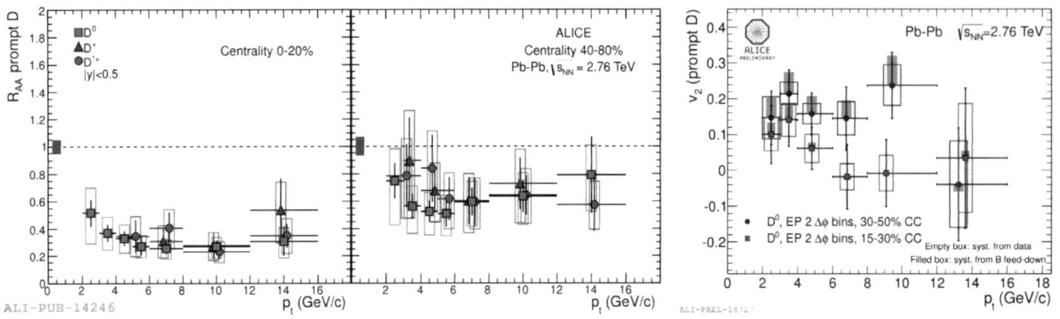

Figure 1. $R_{\rm AA}$ for prompt D^0, D^+ and D^{*+} in the 0-20% (left) and 40%-80% (middle) centrality classes [5]. The statistical (bars), systematic (empty boxes) and normalization (filled boxes) uncertainties are displayed. Right: v_2 of D^0 mesons in the 15%–30% and 30%–50% centrality classes. The statistical (bars), systematic (empty boxes) and B feed-down contribution (filled boxes) uncertainties are drawn. Horizontal error bars represent the bin widths.

2. Heavy-flavour and J/ψ detection in the ALICE experiment

The ALICE experiment, described in detail in [1], is composed of a central barrel (pseudo-rapidity coverage $|\eta| < 0.9$), a muon spectrometer ($-4 < \eta < -2.5$) and a set of detectors for global collision characterization and triggering located in the forward and backward pseudo-rapidity regions. Among those, the VZERO detector, made of two scintillator arrays, and the two Zero Degree Calorimeters (ZDC) are used for the analyses presented here. The Silicon Pixel Detector (SPD), the innermost part of the Inner Tracking System (ITS) is included in the trigger logic and is also needed for the interaction vertex reconstruction.

Open heavy flavours are measured in the charm hadronic decay channels (including $D^0 \rightarrow K^-\pi^+$, $D^+ \rightarrow K^-\pi^+\pi^+$, $D^{*+} \rightarrow D^0\pi^+$, $D_s^+ \rightarrow K^+K^-\pi^+$ and charge conjugates in $|y| < 0.5$), in the semi-electronic decay channel (inclusive $D, B \rightarrow e^{\pm} + X$ in $|y| < 0.8$) and in the semi-muonic decay channel ($D, B \rightarrow \mu^{\pm} + X$ in $2.5 < y < 4$). J/ψ are reconstructed down to $p_{\rm T} = 0$ via unlike-sign dielectrons and unlike-sign dimuons at mid-rapidity ($|y| < 0.9$) and forward rapidity ($2.5 < y < 4$), respectively. The detectors used for particle tracking and identification are the ITS, the Time Projection Chamber (TPC), the Transition Radiation Detector (TRD), the Time of Flight (ToF) and the ElectroMagnetic Calorimeter (EMCal) in the central barrel, and the muon spectrometer at forward rapidity. Details on the reconstruction of D mesons and J/ψ, and on the identification of electrons and muons from heavy-flavour decays can be found in [4–10].

The results are based on the analysis of data samples collected in 2010 and 2011 with minimum bias (MB) and (di)muon triggers. Events are classified according to their centrality by means of the sum of the amplitudes of the signals in the VZERO detectors [11].

3. Heavy-flavour nuclear modification factor and elliptic flow measurements

The study of in-medium effects via the heavy-flavour $R_{\rm AA}$ requires the measurement of the production cross sections in pp collisions (Eq. (1)). The latter are obtained from the pp data sample at $\sqrt{s} = 2.76$ TeV (semi-leptonic channels [10]) or/and the pp data at $\sqrt{s} = 7$ TeV by scaling down to $\sqrt{s} = 2.76$ TeV according to pQCD calculations [12] (hadronic and semi-electronic channels [4, 8]).

Figure 1 shows the $p_{\rm T}$-differential $R_{\rm AA}$ for prompt D^0, D^+ and D^{*+} in the 0–20% (left) and 40%–80% (middle) centrality classes. The results for the three meson species agree within uncertainties. They show a suppression reaching a factor 3–4 in central collisions, for $p_{\rm T} > 5$ GeV/c. For $p_{\rm T} < 5$ GeV/c, the D^0 meson $R_{\rm AA}$ shows a weak rise with decreasing $p_{\rm T}$. The

40%-80% centrality class exhibits less suppression by roughly a factor of two. A complementary analysis to the parton energy loss study by means of the R_{AA} is provided by the elliptic flow measurement. The p_{T}-differential v_2 of D^0 mesons in the centrality classes 15%–30% and 30%–50% is presented in Fig. 1 (right), v_2 being extracted from the relative difference between the number of candidates in-plane and out-of-plane [13]. The results for semi-central collisions (30%–50%) show a non-zero v_2 (with a 3 σ significance) in $2 < p_{\mathrm{T}} < 6$ GeV/c. The magnitude of v_2 decreases as the collision centrality increases.

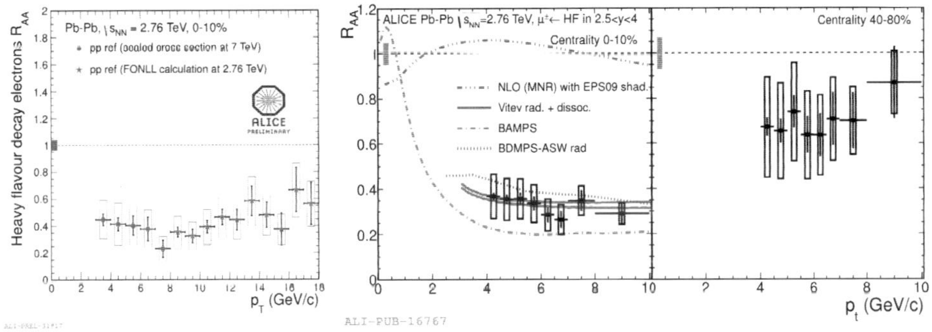

Figure 2. R_{AA} of electrons (left) and muons (middle and right) from heavy-flavour decays for different centrality classes. The statistical (bars), systematic (empty boxes) and normalization (filled boxes) uncertainties are shown. R_{AA} of muons from heavy-flavour decays [10] is compared to models including only shadowing and to models implementing in-medium energy loss.

The p_{T}-differential R_{AA} is also measured for electrons from heavy-flavour decays in the 0–10% centrality class, as displayed in Fig. 2 (left). A suppression of a factor 2–4 is evidenced in $3 < p_{\mathrm{T}} < 18$ GeV/c. The R_{AA} of muons from heavy-flavour decays in the two centrality classes 0–10% and 40%–80% is shown in Fig. 2 (middle and right). A larger suppression is measured in central collisions than in peripheral collisions with no significant p_{T} dependence. The suppression reaches a factor 3–4 in the 10% most central collisions. The comparison with model predictions (Fig. 2, middle) indicates that nuclear shadowing alone cannot explain the observed suppression at forward rapidity [14]. Moreover, models implementing in-medium partonic energy loss [15] describe reasonably well the data within uncertainties. A similar agreement between the average nuclear modification factor of D mesons and transport model predictions was also reported [5].

4. J/ψ nuclear modification factor and elliptic flow measurements

The inclusive J/ψ R_{AA} has been measured at mid-rapidity and forward rapidity, down to $p_{\mathrm{T}} = 0$, in the dielectron channel and in the dimuon channel, respectively. The pp reference (Eq. (1)) is obtained from the analysis of the pp run at $\sqrt{s} = 2.76$ TeV [6]. The inclusive J/ψ R_{AA} as a function of $\langle N_{\mathrm{part}} \rangle$ (mean number of participating nucleons), p_{T} and y is displayed in the left, middle and right panel of Fig. 3, respectively. A clear suppression of about a factor two is measured at forward rapidity for $\langle N_{\mathrm{part}} \rangle > 70$, with almost no centrality dependence. A similar pattern could be evidenced at mid-rapidity within uncertainties. In the forward rapidity region, the J/ψ suppression increases significantly with increasing p_{T} (R_{AA} varies from about 0.6 to 0.35 from $p_{\mathrm{T}} = 0$ up to $p_{\mathrm{T}} = 8$ GeV/c). The suppression measured at high p_{T} in ALICE is similar to that reported by the CMS Collaboration [16], although in a different rapidity region ($1.6 < |y| < 2.4$). On the other hand, in the range $p_{\mathrm{T}} < 4$ GeV/c, a stronger suppression is measured in $1.2 < |y| < 2.2$ for the 20% most central Au–Au collisions at $\sqrt{s_{\mathrm{NN}}} = 0.2$ TeV [17] compared to the ALICE measurement in $2.5 < y < 4$. These observations hint at regeneration as one of the mechanisms responsible for J/ψ production at the LHC. It is worth noticing that

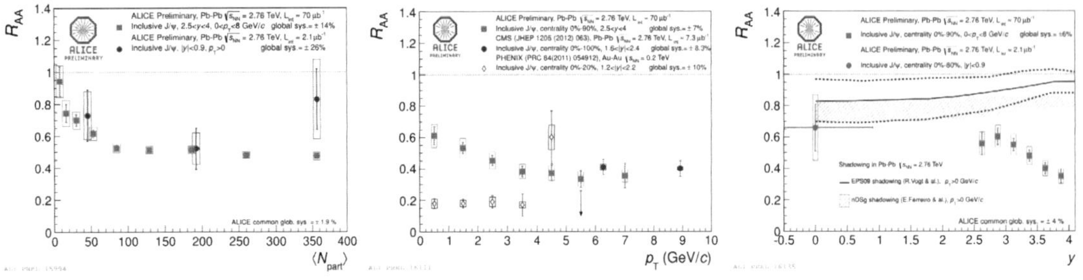

Figure 3. R_{AA} of inclusive J/ψ as function of $\langle N_{part} \rangle$ (left), p_T (middle) and y (right). The p_T-differential R_{AA} from ALICE is compared to the one measured in PHENIX and CMS Collaborations and the y-differential R_{AA} is compared to model calculations including only shadowing effects (EPS09 and nDSg parameterizations [18]).

transport models including a J/ψ regeneration component from deconfined charm quarks in the medium and the statistical hadronization model reproduce the data within uncertainties (see [7] and references therein). The inclusive J/ψ R_{AA} increases by about 40% from $y = 4$ to $y = 2.5$ and tends to saturate to $y = 0$ (Fig. 3). The comparison of the data with model predictions [18] indicates that shadowing effects are expected to be similar over the whole rapidity range and could be responsible for a large part of the measured suppression in $0 < y < 3$. Finally, a non-zero J/ψ v_2 [19] (2.2 σ significance) was recently measured at intermediate p_T in semi-central collisions, supporting that a fraction of J/ψ could be produced by recombination processes

5. Conclusion

The production of open heavy-flavour particles and inclusive J/ψ has been measured at mid-rapidity and forward rapidity in pp and Pb–Pb collisions with the ALICE detector. The open heavy-flavour results give evidence for strong in-medium energy loss of heavy quarks. The inclusive J/ψ results indicate that a significant fraction of J/ψ may be produced from charm quarks in the deconfined medium. The p–Pb run scheduled for the beginning of 2013 should allow us to assess initial state effects and to quantify the amount of nuclear shadowing.

References

[1] Aamodt K *et al.* (ALICE Collaboration) 2008 JINST **3** S08002.
[2] Matsui T and Satz H 1986 Phys. Lett. B **178** 416.
[3] Dokshitzer Y L and Kharzeev D E 2001 Phys. Lett. B **519** 199.
[4] Abelev B *et al.* (ALICE Collaboration) 2012 JHEP **1201** 128.
[5] Abelev B *et al.* (ALICE Collaboration) 2012 JHEP **09** 112.
[6] Abelev B *et al.* (ALICE Collaboration) arXiv:1203.3641 [hep-ex].
[7] Abelev B *et al.* (ALICE Collaboration) 2012 Phys. Rev. Lett. **109** 072301
[8] Abelev B *et al.* (ALICE Collaboration) arXiv:1205.5423 [hep-ex].
[9] Abelev B *et al.* (ALICE Collaboration) 2012 Phys. Lett. B **708** 265.
[10] Abelev B *et al.* (ALICE Collaboration) 2012 Phys. Rev. Lett. **109** 112301.
[11] Aamodt K *et al.* (ALICE Collaboration) 2011 Phys. Rev. Lett. **106** 0320301; 2011 Phys. Lett. B **696** 30.
[12] Averbeck R et al. arXiv:1201.3791 [hep-ph].
[13] Ortona G for the ALICE Collaboration, proceedings of Hard Probes 2012 Conference.
[14] Mangano M L *et al.* 1992 Nucl. Phys. B **373** 295; Eskola K *et al.* 2009 JHEP **0904** 065.
[15] Uphoff J *et al.* arXiv:1205.4945; Armesto N *et al.* 2005 Phys. Rev. D **71** 054027; Sharma R *et al.* 2009 Phys. Rev. C **80** 054902.
[16] Chatrchyan S *et al.* (CMS Collaboration) 2012 JHEP **1205** 063.
[17] Adare A. *et al.* (PHENIX Collaboration) 2011 Phys. Rev. C **84** 054912.
[18] Ferreiro E *et al.* 2011 Nucl. Phys. A **855** 327; Vogt R 2010 Phys. Rev. C **81** 044903.
[19] Suire C for the ALICE Collaboration, arXiv:1208.5601.

Journal of Physics: Conference Series **422** (2013) 012015 doi:10.1088/1742-6596/422/1/012015

Heavy flavor and Quarkonia in heavy-ion collisions with thc CMS

Hyunchul Kim on behalf of the CMS Collaboration
Department of Physics, Korea University, Seoul 136-701, REPUBLIC of KOREA
E-mail: hyunchul.kim@cern.ch

Abstract.
Quarkonia are important probes for diagnosing and characterizing the quark-gluon plasma since they are produced in the early stage and propagate through the medium. The Compact Muon Solenoid (CMS) detector at the LHC is ideally suited to measure muon pairs from quarkonia in the high-multiplicity environment of nucleus-nucleus collisions. During the 2011 heavy-ion run, CMS has recorded an integrated luminosity of about 150 μb^{-1}, which is about a factor of twenty higher compared to the 2010 heavy-ion data. With this significantly larger data sample, CMS has analyzed the nuclear modification factors of prompt J/ψ, non-prompt J/ψ from B mesons, ψ(2S) and separated three upsilon states (1S, 2S, 3S) in PbPb collisions at $\sqrt{s_{NN}}$ = 2.76 TeV.

1. Introduction

A phase transition to a state of quark-gluon plasma (QGP) is predicted by lattice quantum chromodynamics at large energy densities and high temperatures. In the hot and dense medium, the binding potential of the quarkonia, which consist of heavy quark-antiquark pairs such as $c\bar{c}$ or $b\bar{b}$, is expected to be screened by the presence of deconfined color charges. The quarkonia states are then dissociated or are prevented from forming. Because the screening length depends on the temperature, the different quarkonia states that have different binding energy and radius are expected to be dissociated sequentially in the QGP. This sequential suppression is considered as the one of the powerful signatures for existence of deconfined quarks and gluons [1, 2, 3, 4]. Previous SPS and RHIC results show suppression of the J/ψ yield in heavy-ion collisions compared to pp collisions [5, 6, 7, 8], but several observations are yet to be understood. Despite of 10 times higher center-of-mass energy, a similar J/ψ suppression is observed at SPS and RHIC, and the suppression of J/ψ does not increase with local energy density. Heavy-ion collisions at the LHC can help us to resolve those puzzles, as the beam energy and luminosity are more than ten times larger than those at RHIC. Additionally, the CMS detector has excellent capabilities for muon detection and provides measurements of ψ(2S) and the Υ family, which enables the quantitative analysis of quarkonia.

Until now the LHC has provided PbPb collisions twice at $\sqrt{s_{NN}}$ = 2.76 TeV in 2010 and 2011. During the first heavy-ion run of LHC in 2010, the LHC delivered 7.28 μb^{-1} integrated luminosity. In 2011 thanks to the significant improvement of the accelerator performance, CMS recorded about 150 μb^{-1}. In addition, CMS collected 231 nb^{-1} of pp collisions at the same center-of-mass energy and these data serve as a reference for the measurements in PbPb collisions.

The CMS detector, as evident by its name, Compact Muon Solenoid, is optimized for the detection of muons. The muon analysis uses the inner tracker and the muon system. The inner tracking system is located near the collision vertex, and is used to reconstruct the tracks of charged particles, including muons. The muon detectors are located outside of the superconducting solenoid magnet. In the barrel and endcap region, drift tubes and cathode strip chambers are employed for tracking respectively. Resistive plate chambers are used for muon triggering in both regions [9].

2. Results

In order to study effects due to the medium, the nuclear modification factor R_{AA} is analyzed. R_{AA} is defined by the ratio of the yield in nucleus-nucleus (AA) collisions to the yield in pp collisions scaled by the number of average binary nucleon-nucleon collisions:

$$R_{AA} = \frac{\mathcal{L}_{pp}}{T_{AA}N_{MB}} \frac{N_{PbPb}(J/\psi)}{N_{pp}(J/\psi)} \frac{\epsilon_{pp}}{\epsilon_{PbPb}}$$

Here $\mathcal{L}_{pp}$ is the integrated luminosity of the pp data set, T_{AA} is the nuclear overlap function which is calculated from a Glauber model accounting for the nuclear collision geometry, N_{MB} is the number of minimum-bias events sampled by the event selection, and $\epsilon_{pp}/\epsilon_{PbPb}$ is the ratio of the reconstruction efficiency in pp and PbPb collisions [10].

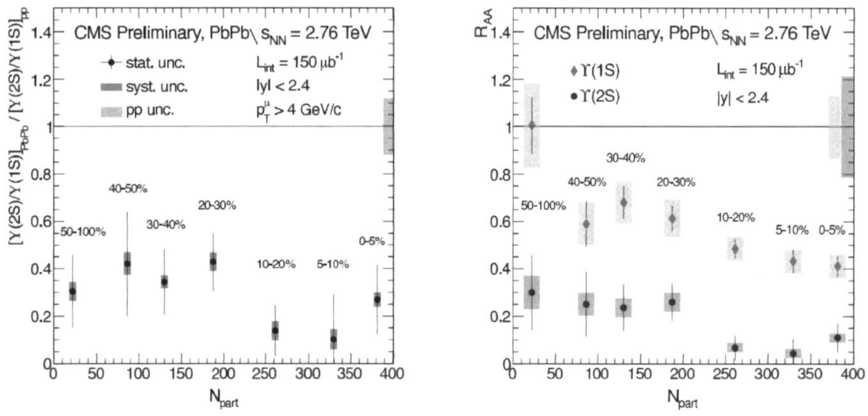

Figure 1. Centrality dependence of double ratio (left) and R_{AA}(right). N_{part} is the number of participating nucleons determined by the Glauber model. (Left): double ratio defined as $(N_{\Upsilon(2S)}/N_{\Upsilon(1S)})_{PbPb}/(N_{\Upsilon(2S)}/N_{\Upsilon(1S)})_{pp}$ versus N_{part}. The global uncertainty was represented by a grey box. (Right): R_{AA} of Υ(1S) (green diamonds) and Υ(2S) (blue circles) as a function of N_{part}. The uncertainties of 14% for Υ(1S) and 21% for Υ(2S) from luminosity, pp yields and efficiency are represented by the boxes at unity and are not included in the data points [12].

Using the first heavy-ion data at LHC for bottomonia analysis, the double ratio of bottomonia yields measured in PbPb and in pp collisions $(N_{\Upsilon(2S+3S)}/N_{\Upsilon(1S)})_{PbPb}/(N_{\Upsilon(2S+3S)}/N_{\Upsilon(1S)})_{pp}$ was estimated to be 0.31 with significance of 2.4 sigma [11]. With the large data sample of heavy-ion data taken in 2011, the Υ(2S) and Υ(3S) states were separated clearly, which enabled us to estimate the double ratio for these two states individually. The left panel of Fig. 1 shows the centrality dependence of the double ratio for the Υ(2S): the centrality integrated double ratio is 0.21 and no strong centrality dependence is observed. For the Υ(3S) because no clear peak can be identified in PbPb collisions, only an upper limit of the double ratio could be estimated as 0.1 with 95% confidence level.

In the right panel of Fig. 1, the centrality dependences of R_{AA} for the Υ(1S) and Υ(2S) states are displayed. The centrality-integrated (0 - 100%) $R_{\rm AA}$ values for each Υ states are $R_{\rm AA}[\Upsilon(1S)] = 0.56 \pm 0.08$ (stat.) $\pm$ 0.07 (syst.), $R_{\rm AA}[\Upsilon(2S)] = 0.12 \pm 0.04$ (stat.) $\pm$ 0.02 (syst.) and $R_{\rm AA}[\Upsilon(3S)] = 0.03 \pm 0.04$ (stat.) $\pm$ 0.01 (syst.). The $R_{\rm AA}$ result for the Υ(3S) state is consistent with the upper limit of 0.1 for the double ratio at 95% confidence level. Assuming about 50% of Υ(1S) originate in the decay of excited states such as Υ(2S) and χ_b, the current result implies the dissociation of only the excited states, not Υ(1S).

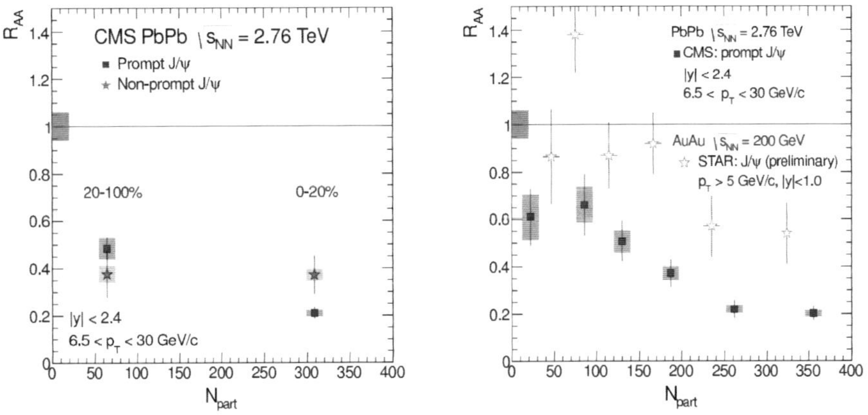

Figure 2. Centrality dependence of $R_{\rm AA}$ for prompt and non-prompt J/ψ, The grey box at $R_{\rm AA} = 1$ represents the global scale uncertainty from the pp data sample. (Left) : Prompt (red squares) and non-prompt (orange stars) J/ψ measured by CMS (Right) : The detailed centrality dependence of $R_{\rm AA}$ for prompt J/ψ compared with the low energy results at 200 GeV obtained by STAR (green stars) [13, 14].

For charmonia analysis, the prompt J/ψ and non-prompt J/ψ from B-meson decay can be separated by using the two-dimensional simultaneous fit of $\mu^+\mu^-$ mass and pseudo-proper decay length which is proportional to the distance between the primary and secondary vertices of J/ψ [13].

The $R_{\rm AA}$ values of prompt and non-prompt J/ψ in measured in the transverse momentum range $6.5 < p_{\rm T} < 30$ GeV/c and in the rapidity interval $|y| < 2.4$ are compared in the left panel of Fig. 2. The results hint the quenching of high-$p_{\rm T}$ b quarks for the first time [13]. The right panel of Fig. 2 shows the centrality dependence of $R_{\rm AA}$ for prompt J/ψ in more detail. In the most central collisions $0 - 10\%$ prompt J/ψ are suppressed by a factor of 5 with respect to the pp reference, and in the most peripheral collisions $50 - 100\%$ they are suppressed by a factor of about 1.6. The right panel of Fig. 2 also shows the $R_{\rm AA}$ of prompt J/ψ measured by STAR for $p_{\rm T} > 5$ GeV/c in AuAu collisions at $\sqrt{s_{NN}} = 200$ GeV. This comparison clearly demonstrates more suppression at LHC than at RHIC [14].

Thanks to the improved statistics in 2011 heavy-ion data, the analysis of ψ(2S) became possible. The double ratio of the yields of ψ(2S) to those of J/ψ measured in PbPb and in pp collisions $(N_{\psi(2S)}/N_{J/\psi})_{\rm PbPb}/(N_{\psi(2S)}/N_{J/\psi})_{\rm pp}$ are analyzed in both low $p_{\rm T}$ ($p_{\rm T} > 3$ GeV/c and $1.6 < |y| < 2.4$) and high $p_{\rm T}$ ($p_{\rm T} > 6.5$ GeV/c and $|y| < 1.6$) and shown in Fig. 3. In the low $p_{\rm T}$ region, ψ(2S) is less suppressed than J/ψ with large uncertainties that are dominated by the limited statistics in the pp collision sample. In the high $p_{\rm T}$ region, ψ(2S) is more suppressed than J/ψ. From the above double ratio the $R_{\rm AA}$ of ψ(2S) were

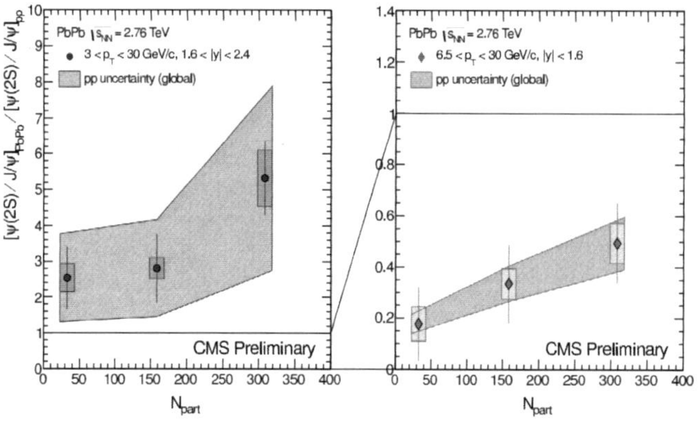

Figure 3. Centrality dependence of double ratio $(N_{\psi(2S)}/N_{J/\psi})_{\rm PbPb}/(N_{\psi(2S)}/N_{J/\psi})_{\rm pp}$. Grey area is global pp uncertainty. (Left) : double ratio in a low $p_{\rm T}$ region, $3 < p_{\rm T} < 30$ GeV/c and $1.6 < |y| < 2.4$ (Right) : double ratio in a high $p_{\rm T}$ region, $6.5 < p_{\rm T} < 30$ GeV/c and $|y| < 1.6$

estimated as 1.54 ± 0.32 (stat.) $\pm$ 0.22 (syst.) $\pm$ 0.76 (pp) in the low $p_{\rm T}$ region, and 0.11 ± 0.03 (stat.) $\pm$ 0.02 (syst.) $\pm$ 0.02 (pp) in the high $p_{\rm T}$ region [15].

3. Summary

With improved statistics of 2011 heavy-ion data, CMS measured the Υ(2S) suppression for the first time and set the upper limit on Υ(3S) production using a double ratio of the yield with respect to the Υ(1S) state measured in PbPb and in pp collisions. The nuclear modification factors $R_{\rm AA}$ for Υ(1S), Υ(2S) and Υ(3S) show suppression by factors of about 2, 8 and lager than 10. These measurements imply the dissociation of the excited states in the Upsilon family, but not the Υ(1S) state. The ψ(2S) are more suppressed than J/ψ in the high $p_{\rm T}$ region, but less suppressed in the low $p_{\rm T}$ region. Future precise measurements in pp collisions are required to draw a definite conclusion in the low $p_{\rm T}$ region.

4. Acknowledgements

This work was supported in part by the National Research Foundation of Korea under grant No. K20802011718-11B1301-00610.

References

[1] T.Matsui and H.Satz 1986 *Phys. Lett.* B **178** 416
[2] H.Satz 2007 *Nucl. Phys* A **783** 249
[3] H.Satz 2006 *J. Phys.* G **32** R25
[4] A.Mocsy 2009 *Eur. Phys. J.* C **61** 705
[5] NA50 Collaboration 2005 *Eur. Phys. J.* C **39** 335
[6] NA60 Collaboration 2007 *Phys. Rev. Lett.* **99** 132302
[7] NA38 Collaboration 1999 *Phys. Lett.* B **449** 128
[8] PHENIX Collaboration 2011 *Phys. Rev.* C **84** 054912
[9] CMS Collaboration 2008 *JINST* **3** S08004
[10] B. Alver *et al.* 2008 *Phys. Rev.* C **77** 014906
[11] CMS Collaboration 2011 *Phys. Rev. Lett.* **107** 052302
[12] CMS Collaboration 2012 accepted by *Phys. Rev. Lett.* arXiv : 1208.2826
[13] CMS Collaboration 2012 *JHEP* **1205** 063
[14] STAR Collaboration 2011 *J. Phys.* G **38** 124007
[15] CMS Collaboration 2012 CMS-PAS-HIN-12-007

Journal of Physics: Conference Series **422** (2013) 012016 doi:10.1088/1742-6596/422/1/012016

Impact of gluon damping on heavy-quark quenching

M. Nahrgang(1,2), **M. Bluhm**(3), **P. B. Gossiaux**(1), **J. Aichelin**(1)

(1) SUBATECH, Université de Nantes, EMN, IN2P3/CNRS, 4 rue Alfred Kastler, 44307 Nantes cedex 3, France

(2) Frankfurt Institute for Advanced Studies (FIAS), Ruth-Moufang-Str. 1, 60438 Frankfurt am Main, Germany

(3) Dipartimento di Fisica, Università degli Studi di Torino and INFN, Sezione di Torino, via Giuria 1, 10125 Torino, Italy

E-mail: nahrgang@subatech.in2p3.fr

Abstract. In this conference contribution, we discuss the influence of gluon-bremsstrahlung damping in hot, absorptive QCD matter on the heavy-quark radiation spectra. Within our Monte-Carlo implementation for the description of the heavy-quark in-medium propagation we demonstrate that as a consequence of gluon damping the quenching of heavy quarks becomes significantly affected at higher transverse momenta.

1. Introduction

The fundamental properties of strongly interacting matter are investigated in ultra-relativistic heavy-ion collisions. With these experiments one aims at understanding the basic processes in QCD matter at finite temperature T and/or baryon density. For this purpose heavy quarks are a particularly clean probe of the formed matter because they are predominantly produced from initial hard scatterings of participating nucleons. Subsequently they undergo collisional and radiative processes through interactions with the locally thermalized light quarks and gluons in the surrounding medium.

These interactions with the medium constituents lead to a thermalization of heavy quarks with small transverse momentum p_T, while those with a large p_T suffer from a significant in-medium energy loss. This is reflected in sensitive observables such as the nuclear modification factor $R_{\rm AA}$. Results from RHIC for the heavy-flavour decay non-photonic single electron $R_{\rm AA}$ as well as the $R_{\rm AA}$ of D-mesons [1] indicate indeed a substantial quenching of heavy quarks at higher p_T up to 10 GeV. The data from the latest measurements at the LHC [2] was able to extend this p_T-range up to 30 GeV for very central collisions. This offers a new and unique opportunity for testing our understanding of in-medium energy loss mechanisms, with the prospect of disentangling eventually the different contributions from collisional and radiative energy loss processes.

Radiative processes are considered to be the dominant contribution to parton energy loss at larger energies, but they may become affected by coherence effects in the medium, cf. the review in [3]. As a consequence of the destructive interference of radiation amplitudes from successive scatterings off medium constituents within the formation length l_f of gluon-bremsstrahlung the radiation spectrum is suppressed from independent scatterings for larger gluon energies ω. This is the QCD-analog of the Landau-Pomerantschuk-Migdal effect [4]. Other effects may alter the radiation pattern as well. For example, the dielectric polarization of the medium modifies the dispersion relation of radiated gluons which aquire an effective in-medium mass m_g. This leads to a significant reduction of the radiation spectrum in the soft-ω region [5, 6]. Similarly, dissipative effects in the QCD medium may diminish the radiative energy loss of heavy quarks as advocated in [7].

In these proceedings, we discuss the dissipative effect of gluon damping on the radiation spectra of heavy quarks. Focussing on the implementation of the novel effect into our Monte-Carlo approach to the heavy-quark in-medium propagation [8, 9] we show that damping mechanisms influence the R_{AA} of heavy quarks in a pronounced way.

2. Gluon damping phenomenology

In an absorptive plasma, mechanisms, which lead to the damping of bremsstrahlung gluons, can reduce the radiative energy loss of traversing, highly-energetic charges. This phenomenon was first demonstrated in [10] for asymptotic electric charges in electro-magnetic plasmas by modelling the dispersive medium via a complex index of refraction and then advocated in [7] to be of relevance also in the absorptive Quark-Gluon Plasma (QGP). In fact, dissipative effects such as quark–anti-quark pair creation or secondary gluon-bremsstrahlung generation introduce an additional scale in the medium. If this damping length scale l_d, which is related to the gluon damping rate Γ as $l_d \simeq 1/\Gamma$, is smaller than or of the order of l_f of the nascent gluon, absorptive mechanisms will reduce the probability for the formation of gluon-bremsstrahlung and, thus, influence the radiative energy loss of partons. This might be particularly important in the case of the large formation lengths in coherent emission processes.

Even though the aforementioned mechanisms represent higher-order effects in perturbative QCD, they might be relevant for the matter investigated in the laboratory. For example, the damping rate due to secondary gluon-bremsstrahlung formation is parametrically given as $\Gamma \sim g^4 T$ up to logarithmic corrections of order $\mathcal{O}(1/g)$ [7]. Consequently, higher temperatures imply a larger damping rate and thus a smaller l_d. Moreover, the formation length of gluon radiation at fixed $x = \omega/E$ increases with increasing energy E of the emitting parton [7]. Thus, with increasing E and/or increasing T dissipative effects become more and more important for the radiation spectra of partons, a situation that is likely to be encountered experimentally at the LHC.

3. Medium-modification of the heavy-quark radiation spectrum

The Gunion-Bertsch (GB) gluon radiation spectrum [11] originating from single, independent scatterings of massless partons was generalized within scalar QCD to the case of massive quarks in [12]. Medium-modifications of this heavy-quark GB-spectrum due to coherence effects were discussed in detail in [13]. In Fig. 1, we show the radiation spectra of both cases for charm and bottom quarks at different parton energies E. (The spectra which show the influence of coherence effects are labelled "LPM".)

In order to highlight quantitatively the additional impact of gluon damping effects on the bremsstrahlung spectrum off heavy quarks, we make use of a scaling ansatz advocated in [14] according to which the global radiation intensity becomes modified due to medium effects via $d^2I = d^2I_{GB} \cdot \tilde{l}_f / l_f^0$. Here, the scale $\tilde{l}_f = \min(l_d, l_f)$ is given by the minimum between damping length l_d and formation length l_f discussed in [7], whereas $l_f^0 \simeq 2x(1-x)E/(m_g^2 + x^2 m_s^2)$ with the parton mass m_s denotes the formation length in a single, independent scattering process, cf. [7, 13]. Figure 1 exhibits the correspondingly modified spectra for two different values of Γ. As evident, damping effects may reduce the heavy-quark radiation spectrum significantly in an intermediate ω-region, where the influence of the dissipative effects increases with both increasing Γ and E. This reduction is stronger than the reduction due to the LPM effect. Moreover, with larger Γ and larger E the spectra become more and more quark-mass independent.

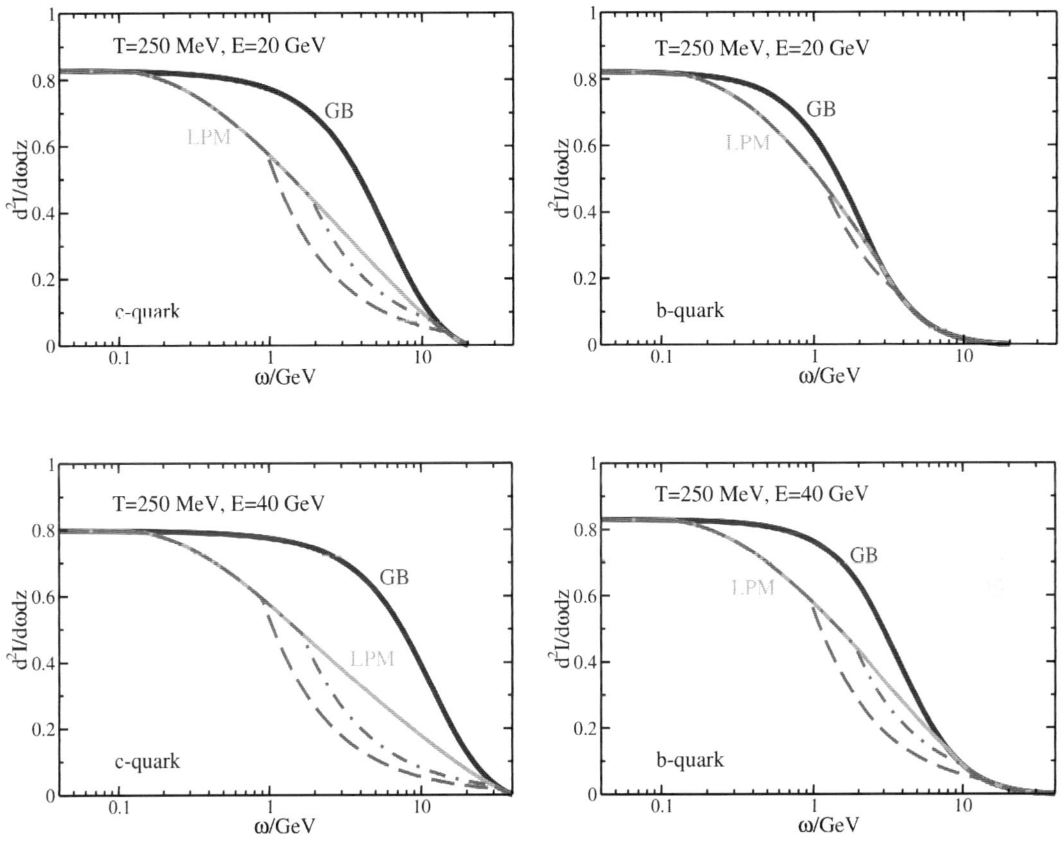

Figure 1. (Colour online) Suppression of the heavy-quark Gunion-Bertsch (GB) radiation spectrum (thick solid curves) due to LPM-effect (thinner solid curves) and gluon damping effect (dash-dotted curves for a damping rate $\Gamma = 0.5\,T$ and dashed curves for $\Gamma = 0.75\,T$, with the fixed gluon mass of $m_g = 2\,T$). Left panels are for charm quarks with $m_c = 1.5$ GeV, right panels for bottom quarks with $m_b = 5.1$ GeV, while the upper (lower) row shows the results for a quark energy E of 20 GeV (40 GeV).

4. Influence of gluon damping on the quenching of heavy quarks

To study the consequences of gluon damping on observables, we include the modified radiation spectra illustrated above in the Monte-Carlo approach [8, 9] to the heavy-quark propagation in the medium. The interaction of the heavy quarks with the light partons in the QGP amounts in collisional and radiative processes, which are described in detail in [8, 9, 12, 13], while the evolution of the medium is obtained from the fluid dynamic expansion presented in [15]. Any theoretical uncertainty affecting the model is assumed to be captured in one global factor K which rescales the interaction rates and which is obtained from calibration to the available RHIC data. In this spirit it is possible to reproduce the available heavy-flavour data at RHIC energies for both purely collisional and collisional plus radiative (without damping) energy loss scenarios as was demonstrated in [8, 12, 13]. At LHC energies, the p_T-range is extended and we are able to identify different trends in the R_{AA} for these two scenarios, cf. [16]. Counter-intuitively, available D-meson data from central $Pb+Pb$ collisions at the LHC [2] seem to favour the purely collisional energy loss scenario [16]. This indicates that a reduction of the radiative energy loss component would be necessary in order to reconcile the model with the data. As illustrated in Fig. 2, one possibility for such a reduction could be the inclusion of gluon damping mechanisms.

Here, we show the $R_{\rm AA}$ of charm and bottom quarks for the collisional plus radiative energy loss scenario (dashed curves), which includes the LPM effect but no dissipative effects. The inclusion of gluon damping mechanisms results in the solid curves. As evident, when one neglects gluon damping effects, the $R_{\rm AA}$ stays almost flat for $p_T \geq 5$ GeV for the charm quarks and for $p_T \geq 15$ GeV for the heavier bottom quarks. Once gluon damping is taken into account, the nuclear modification factor increases visibly at higher transverse momenta. In the shown p_T-range predominantly charm quarks are affected, while the effect sets in for bottom quarks only at larger p_T. Details, however, depend on the value of the gluon damping rate.

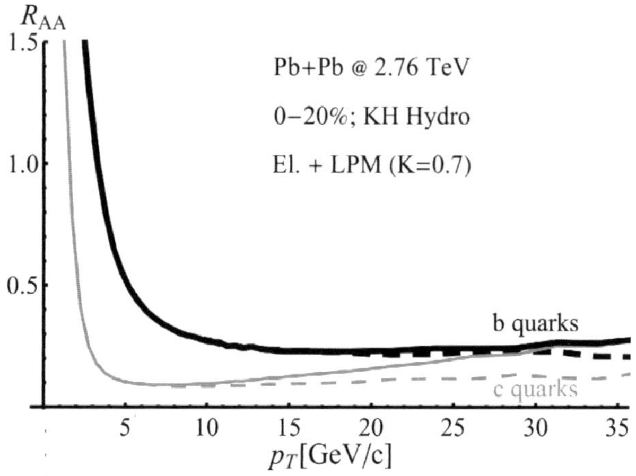

Figure 2. (Colour online) $R_{\rm AA}$ of charm (thin curves) and bottom quarks (thick curves) with (solid) and without (dashed) the effect of gluon damping. The damping rate is $\Gamma = 0.75\,T$. For details cf. [16].

5. Conclusions

We discussed the influence of gluon damping in the absorptive QCD medium on the quenching of heavy quarks in ultra-relativistic heavy-ion collisions. In addition to the known modification of radiation spectra due to coherence effects, gluon damping effects reduce significantly the spectra in an intermediate ω-region for large gluon damping rates Γ and/or large parton energies E. Consequently, in the absorptive QGP heavy quarks become less quenched at large transverse momenta. With increasing p_T, the quenching seems to become quark-mass independent which would be visible in a prominent behaviour of the heavy-flavour meson $R_{\rm AA}$ as advocated in [16].

MN and MB thank the organisers of the conference *Heavy Ion Collisions in the LHC Era, Quy Nhon, Vietnam, 16-20 July 2012*, for financing their participation. MB acknowledges the financial support received from the SaporeGravis Network of Hadron Physics I3 (I3HP3). MN thanks her parents for financing her flight to Vietnam.

References

[1] STAR Collab., Phys. Rev. Lett. **98** (2007) 192301; Erratum-ibid. **106** (2011) 159902: PHENIX Collab., Phys. Rev. C **84** (2011) 044905; X. Dong [for the STAR Collaboration], arXiv:1210.6677 [nucl-ex].
[2] ALICE Collab., JHEP **09** (2012) 112.
[3] S. Peigné and A. V. Smilga, Phys. Usp. **52** (2009) 659.
[4] R. Baier, Y. Dokshitzer, A. Müller, S. Peigné, D. Schiff, Nucl. Phys. B **483** (1997) 291; ibid. **484** (1997) 265.
[5] B. Kämpfer and O. P. Pavlenko, Phys. Lett. B **477** (2000) 171.
[6] M. Djordjevic and M. Gyulassy, Phys. Rev. C **68** (2003) 034914; Phys. Lett. B **560** (2003) 37.
[7] M. Bluhm, P. B. Gossiaux, T. Gousset, J. Aichelin, arXiv:1204.2469 [hep-ph]; arXiv:1209.1149 [hep-ph].
[8] P. B. Gossiaux and J. Aichelin, Phys. Rev. C **78** (2008) 014904.
[9] P. B. Gossiaux, R. Bierkandt and J. Aichelin, Phys. Rev. C **79** (2009) 044906.
[10] M. Bluhm, P. B. Gossiaux, J. Aichelin, Phys. Rev. Lett. **107** (2011) 265004; J. Phys. G **38** (2011) 124119.
[11] J. F. Gunion and G. Bertsch, Phys. Rev. D **25** (1982) 746.
[12] P. B. Gossiaux, J. Aichelin, T. Gousset and V. Guiho, J. Phys. G **37** (2010) 094019.
[13] P. B. Gossiaux, arXiv:1209.0844 [hep-ph].
[14] V. M. Galitsky and I. I. Gurevich, Il Nuovo Cimento **32** (1964) 396.
[15] P. F. Kolb, J. Sollfrank and U. Heinz, Phys. Rev. C **62** (2000) 054909
[16] P. B. Gossiaux, M. Nahrgang, M. Bluhm, T. Gousset and J. Aichelin, Quark Matter 2012 proceedings.

Recent Heavy Flavour Results from ATLAS

Marcella Bona, on behalf of the ATLAS Collaboration

School of Physics and Astronomy, Queen Mary, University of London, Mile End Road, London E1 4NS, UK

E-mail: m.bona@qmul.ac.uk

Abstract. We present here a selection of heavy flavour physics results from the ATLAS experiment taking data at the Large Hadron Collider. The ATLAS analyses allowed a variety of heavy flavour measurements, reproducing essential B-hadron properties and demonstrating a good performance of the detector. We include in this report the measurement of the centrality dependence of J/ψ yields in lead-lead collisions, observations of χ_{cJ} and χ_b states with a first observation of a $\chi_b(3P)$ state, the search for the rare decay $B_s^0 \to \mu^+\mu^-$ and, finally, a measurement of $B_s \to J/\psi\phi$ decay parameters.

1. Introduction

The ATLAS (A Toroidal LHC ApparatuS) experiment [1] is a general purpose experiment operating at the Large Hadron Collider (LHC) at CERN. It has a rich heavy flavour program including measurements of b-quark production, studies of b-hadron decays, as well as measurements of quarkonium and exotic states production. ATLAS is now performing also indirect searches for new physics, such as the rare decays of $B_s^0 \to \mu^+\mu^-$, and measurements of CP-violating phase in the $B_s^0 \to J/\psi\phi$ decay. These provide important constraints to the Standard Model (SM) and are complementary to direct searches for new physics.

2. Measurement of the centrality dependence of J/ψ yields

The centrality dependence is studied in the yield of J/ψ mesons produced in the collisions of lead ions at the LHC [2, 3, 4]. In a sample of minimum-bias lead-lead collisions at a nucleon-nucleon centre of mass energy $\sqrt{s_{NN}} = 2.76$ TeV, corresponding to an integrated luminosity of about 6.7 μb^{-1}, J/ψ mesons are reconstructed via their decays to $\mu^+\mu^-$ pairs. In ATLAS, the J/ψ yields are normalised to the $40-80\%$ centrality bin (peripheral) to obtain the "relative yield". These are then compared to R_{coll}, where R_{coll} is the ratio of the mean number of binary collisions in each centrality bin to that of the most peripheral ($40-80\%$) bin.

A clear difference is observed as a function of centrality between the measured relative J/ψ yield and the prediction based on R_{coll}, indicating a deviation from the simplest expectation based on elementary nucleon-nucleon collision scaling. The ratio of these two values, R_{cp}, is shown as a function of centrality in Figure 1. The data points are not consistent with their average value; instead, a significant decrease of R_{cp} as a function of centrality is observed.

3. Heavy quarkonium physics

Quarkonia are formed from a quark-antiquark pair of same flavour and should represent one of the simplest systems described by QCD theory. Heavy quarkonium allows for rigorous tests

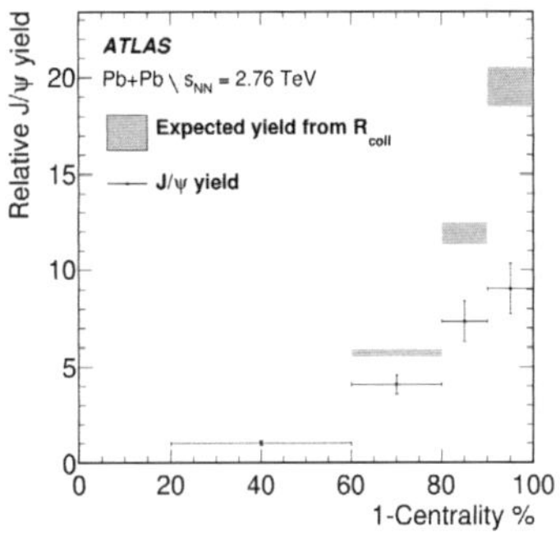

Figure 1. Relative J/ψ yield as a function of centrality normalised to the most peripheral bin (black dots) [3]. The expected relative yields are shown (boxes) with 1σ systematic uncertainties).

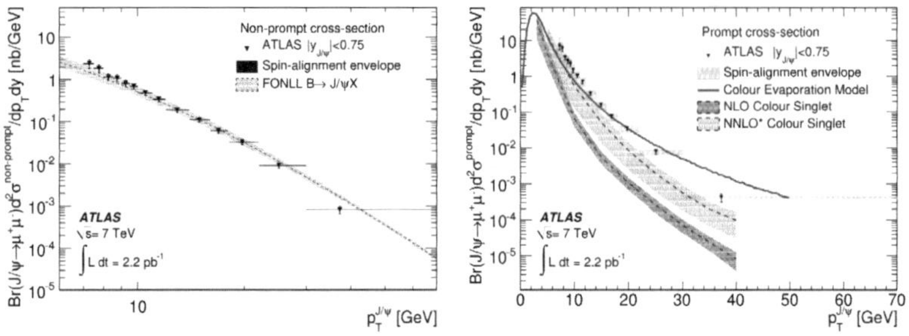

Figure 2. *Left*: Non-prompt J/ψ production cross-section as a function of J/ψ transverse momentum, compared to predictions from FONLL theory in the central rapidity bin [5]. *Right*: Prompt J/ψ production cross-section as a function of J/ψ transverse momentum in the central rapidity bin [5]. Predictions from NLO and NNLO* calculations, and from CEM are overlaid.

of the interplay between perturbative and non-perturbative QCD and a rich spectrum of radial and orbital excitations allow studies of spectroscopy and decay dynamics [6].

3.1. Charmonium: J/ψ Differential Production Cross-section

Using 2.2 pb^{-1} of data from 2010, ATLAS has measured the inclusive $J/\psi \to \mu^+\mu^-$ cross-section and the fraction of J/ψ which are produced non-promptly via decay of a B-hadron [5]. By combining these two measurements, separate cross-sections for the prompt and non-prompt J/ψ's are also made.

Figure 2 shows the central rapidity values of the non-prompt and prompt J/ψ production cross-sections as a function of J/ψ transverse momentum compared to theoretical predictions. Overlaid on the plots is a band representing the variation of the result under various spin-alignment scenarios representing a theoretical uncertainty on the prompt component. The measured non-prompt cross-section is in good agreement with Fixed-Order Next-to-Leading-Log (FONLL) theoretical predictions. The prompt cross-section is compared to Colour Singlet (CSM) next-to-leading order (NLO) and partial next-to-next-leading order (NNLO*) predictions and to the phenomenological Colour Evaporation Model (CEM).

3.2. Charmonium Spectroscopy: Observation of χ_{cJ} (1P) State

ATLAS has performed the observation of the $\chi_{c1}(1P)$ and $\chi_{c2}(1P)$ charmonium states in $\chi_c \to J/\psi\gamma$ decays using 39 pb^{-1} of 2010 pp collision data. [7]. J/ψ candidates are reconstructed from the decay $J/\psi \to \mu^+\mu^-$ while photons are reconstructed using calorimetric measurements. An extended unbinned maximum likelihood fit is performed to the invariant mass difference of the $\mu^+\mu^-$ and $\mu^+\mu^-\gamma$ systems to yield a total of 2960 ± 120(stat.)± 90(syst.) χ_{c1} and χ_{c2} candidates (see Figure 3).

3.3. Bottomonium Spectroscopy: Observation of χ_b state

The P-wave $b\bar{b}$ χ_b states can be reconstructed in ATLAS through the radiative decay to Υ. The $\chi_b(1P)$ and $\chi_b(2P)$ states have already been observed through this decay mode. Using 4.4 fb^{-1} of data from 2011, ATLAS has made the first observation of the $\chi_b(3P)$ state [8].

Photons are reconstructed either directly in the calorimeter or through a conversion to e^+e^-. Figure 4 shows the mass distribution for unconverted photons and converted photons, where the

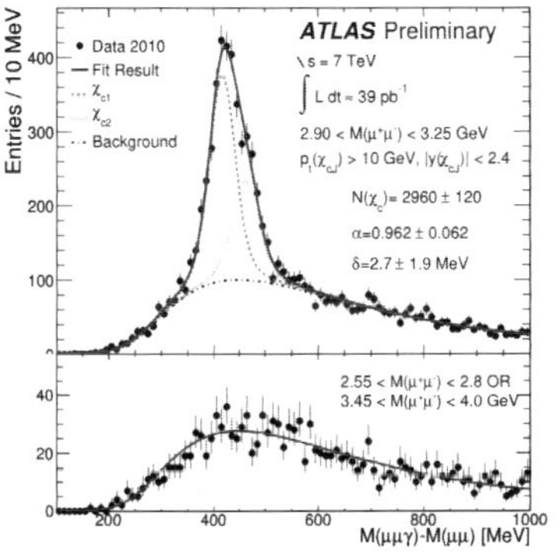

Figure 3. The result of a simultaneous fit to the signal sample (top) and background (J/ψ sideband) sample (bottom) [7]. The individual signal components follow the colour convention: χ_{c1} (red dashed line), and χ_{c2} (magenta dashed line).

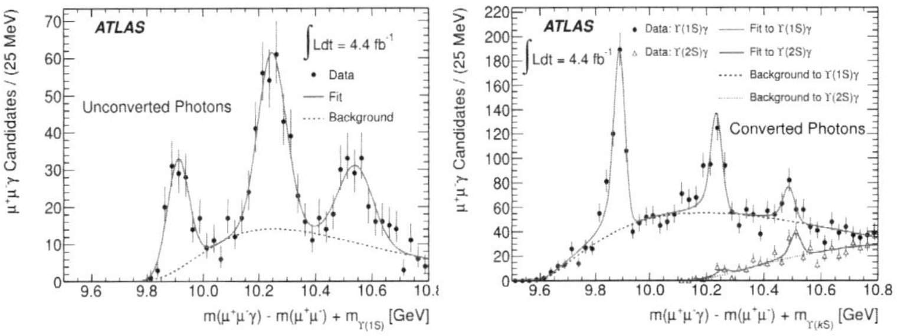

Figure 4. *Left*: The mass distribution of $\chi_b \to \Upsilon(1S)\gamma$ candidates for unconverted photons from energy deposits in the electromagnetic calorimeter [8]. *Right*: The mass distributions of $\chi_b \to \Upsilon(kS)\gamma$ (k = 1, 2) candidates formed using converted photons reconstructed in the inner detector [8]. Data are shown before the correction for the energy loss of the photon conversion electrons. The data for decays of $\chi_b \to \Upsilon(1S)\gamma$ (circles) and $\chi_b \to \Upsilon(2S)\gamma$ (triangles) are plotted. Solid lines represent the total fit result for each mass window. The dashed lines represent the background components only.

mass is defined as $\tilde{m}_k = m(\mu\mu\gamma) - m(\mu\mu) + m_{\Upsilon(kS)}$ and $m_{\Upsilon(kS)}$ are the world average masses of the $\Upsilon(kS)$ states [9]. In addition to the expected peaks for $\chi_b(2P, 1P) \to \Upsilon(2S, 1S)\gamma$, structures are observed at an invariant mass of approximately 10.5 GeV. These additional structures are interpreted as the radiative decays of the previously unobserved $\chi_b(3P)$ states, $\chi_b(3P) \to \Upsilon(2S)\gamma$ and $\chi_b(3P) \to \Upsilon(1S)\gamma$. The measurement with converted photons is used for the final mass determination of 10.539 ± 0.005(stat.)± 0.009(syst.) GeV.

4. B-hadron Masses and Lifetimes

In ATLAS, B-hadrons can be reconstructed exclusively from their decays to J/ψ, *i.e.* $B \to J/\psi(\mu^+\mu^-)X$. A number of B-hadrons have been observed in ATLAS through such decays [10, 11, 12, 13, 14, 15, 16] and masses and lifetimes have been measured showing consistency with PDG values [9].

5. Rare B Decay $B_s^0 \to \mu^+\mu^-$

The rare decays $B_s^0 \to \mu^+\mu^-$ cannot proceed at tree level via flavour-changing neutral-current processes as they are forbidden in the SM and can only occur via higher order diagrams. These decays are also helicity suppressed, resulting in expected SM branching ratios (BR) of $(3.2 \pm 0.2) \times 10^{-9}$ [17]. A larger branching fraction would potentially indicate contributions from New Physics. Until now neither an enhancement of the branching fraction has been observed nor the theoretical limit has been reached experimentally [18, 19, 20, 21].

The first search for $B_s^0 \to \mu^+\mu^-$ in ATLAS is using 2.4 fb $^{-1}$ of the 2011 collision data [22]. ATLAS obtains an observed limit of 2.2×10^{-8} at 95% CL, while the expected limit obtained with the background only hypothesis is $2.3^{+1.0}_{-0.5} \times 10^{-8}$ at 95% CL. These limits are shown in Figure 5.

6. $\Delta\Gamma_s$ and ϕ_s Measurement from $B_s^0 \to J/\psi\phi$

A measurement of $B_s \to J/\psi\phi$ decay parameters, including the CP-violating weak phase ϕ_s and the decay width difference $\Delta\Gamma_s$ is reported, using 4.9 fb^{-1} of 2011 collision data [23]. The mean decay width Γ_s and the transversity amplitudes $|A_0(0)|^2$ and $|A_{\parallel}(0)|^2$ are also measured.

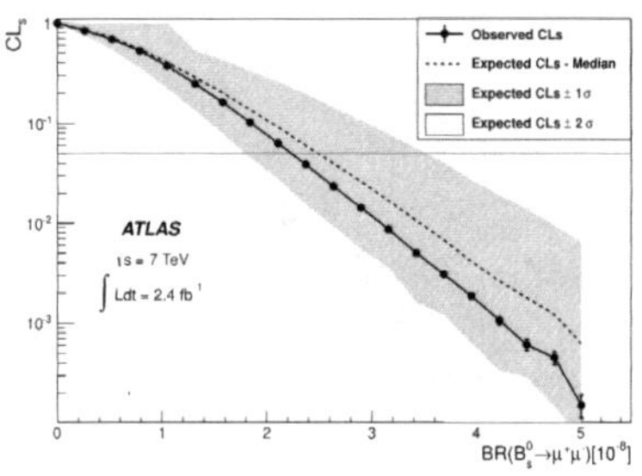

Figure 5. Observed $\mathrm{CL_s}$ (circles) as a function of $\mathrm{BR}(B_s^0 \to \mu^+\mu^-)$ [22]. The 95% CL limit is indicated by the horizontal (red) line. The dark (green) and light (yellow) bands correspond to $\pm 1\sigma$ and $\pm 2\sigma$ fluctuations on the expectation (dashed line).

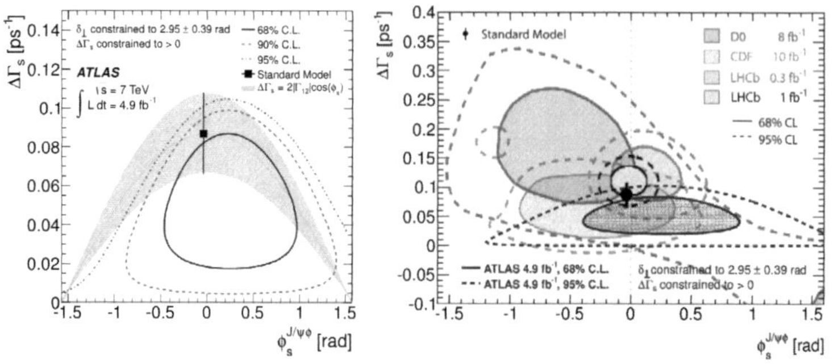

Figure 6. *Left*: Likelihood contours in the $\phi_S - \Delta\Gamma_S$ plane [23]. Three contours show the 68%, 90% and 95% confidence intervals (statistical errors only). The green band is the theoretical prediction of mixing-induced CP violation. *Right*: Likelihood contours in the $\phi_S - \Delta\Gamma_S$ plane [23]. The 68% (solid lines) and 95% (dashed lines) contours are based on statistical errors only. The single black point corresponds to the SM theory value.

Results for ϕ_s and $\Delta\Gamma_s$ are presented in Figure 6, which show agreement with SM expectations. The measured values are: $\phi_s = 0.22 \pm 0.41(\text{stat.}) \pm 0.10(\text{syst.})$ rad $\Delta\Gamma_s = 0.053 \pm 0.021(\text{stat.}) \pm 0.008(\text{syst.})$ ps^{-1} where these correspond to the solution compatible with the external measurements to which the strong phase $\delta_\perp$ is constrained to the LHCb value of 2.95 ± 0.39 rad [24] and where $\Delta\Gamma_s$ is constrained to be positive following the LHCb measurement [25].

References

[1] ATLAS Collaboration 2008 *JINST* **3** S08003
[2] Abelev B *et al.* (ALICE Collaboration) 2012 *Phys.Rev.Lett.* **109** 072301 (*Preprint* 1202.1383)
[3] ATLAS Collaboration 2011 *Phys.Lett.* **B697** 294–312
[4] CMS Collaboration 2012 *JHEP* **1205** 063 (*Preprint* 1201.5069)
[5] ATLAS Collaboration 2011 *Nucl.Phys.* **B850** 387–444 *and references therein.*
[6] Brambilla N, Eidelman S, Heltsley B, Vogt R, Bodwin G *et al.* 2011 *Eur.Phys.J.* **C71** 1534
[7] ATLAS Collaboration, ATLAS-CONF-2011-136, August 2011 URL https://cdsweb.cern.ch/record/1383839
[8] ATLAS Collaboration 2012 *Phys. Rev. Lett.* **108** 152001
[9] Beringer J *et al.* 2012 *Phys.Rev.* **D86** 010001
[10] ATLAS Collaboration, ATLAS-CONF-2010-098, Nov. 2010 URL https://cdsweb.cern.ch/record/1307530
[11] ATLAS Collaboration, ATLAS-CONF-2011-050, April 2011 URL https://cdsweb.cern.ch/record/1341815
[12] ATLAS Collaboration, ATLAS-CONF-2011-105, July 2011 URL https://cdsweb.cern.ch/record/1369830
[13] ATLAS Collaboration, ATLAS-CONF-2011-124, August 2011 URL https://cdsweb.cern.ch/record/1378514
[14] ATLAS Collaboration, ATLAS-CONF-2012-028, March 2012 URL https://cdsweb.cern.ch/record/1430737
[15] ATLAS Collaboration, ATLAS-CONF-2011-092, July 2011 URL https://cdsweb.cern.ch/record/1363779
[16] ATLAS Collaboration 2012 Submitted to Phys. Rev. D URL http://arxiv.org/abs/1207.2284
[17] Buras A J, Carlucci M V, Gori S and Isidori G 2010 *JHEP* **1010** 009
[18] D0 Collaboration 2010 *Phys.Lett.* **B693** 539–544
[19] CDF Collaboration 2011 *Phys.Rev.Lett.* **107** 239903
[20] CMS Collaboration 2012 *JHEP* **1204** 033
[21] LHCb collaboration 2012 *Phys.Rev.Lett.* **108** 231801
[22] ATLAS Collaboration 2012 *Phys.Lett.* **B713** 387–407
[23] ATLAS Collaboration 2012 (*Preprint* 1208.0572) URL http://arxiv.org/abs/1208.0572
[24] LHCb collaboration 2012 *Phys.Rev.Lett.* **108** 101803
[25] LHCb collaboration 2012 *Phys.Rev.Lett.* **108** 241801

Journal of Physics: Conference Series **422** (2013) 012018 doi:10.1088/1742-6596/422/1/012018

J/ψ and ψ' production in proton(deuteron)-nucleus collisions: lessons from RHIC for the proton-lead LHC run

E. G. Ferreiro[a], **F. Fleuret**[b], **J.P. Lansberg**[c] **and A. Rakotozafindrabe**[d]

[a]Departamento de Física de Partículas and IGFAE, Universidad de Santiago de Compostela, 15782 Santiago de Compostela, Spain

[b]Laboratoire Leprince Ringuet, École Polytechnique, CNRS/IN2P3, 91128 Palaiseau, France

[c]IPNO, Université Paris-Sud, CNRS/IN2P3, F-91406, Orsay, France

[d]IRFU/SPhN, CEA Saclay, 91191 Gif-sur-Yvette Cedex, France

E-mail: elena@fpaxp1.usc.es

Abstract. We study the impact of different cold nuclear matter effects both on J/ψ and ψ' production, among them the modification of the gluon distribution in bound nucleons, commonly known as *gluon shadowing*, and the survival probability for a bound state to escape the nucleus –*the nuclear absorption*. Less conventional effects such as *saturation* and *fractional energy loss* are also discussed. We pay a particular attention to the recent PHENIX preliminary data on ψ' production in dAu collisions at $\sqrt{s} = 200$ GeV, which show a strong suppression for central collisions, 5 times larger than the one obtained for J/ψ production at the same energy. We conclude that none of the abovementioned mechanisms can explain this experimental result.

1. Introduction

The properties of the production and the absorption of quarkonium in proton(deuteron)-nucleus collisions, such as J/ψ and ψ' suppression in dAu collisions at RHIC, offer the opportunity to study Cold Nuclear Matter (CNM) effects in the nuclear medium. This is of particular relevance for the understanding of QCD at high density and temperature in nucleus-nucleus collisions, since it is in fact adventurous to claim that the charmonium family is a quark-gluon plasma (QGP) thermometer [1] without calibrating first the CNM suppression.

We have studied the impact of different cold nuclear matter effects both on J/ψ and ψ' production, namely the survival probability for a bound state to escape the nucleus –*the nuclear absorption*– and the modification of the gluon distribution in bound nucleons –*gluon shadowing*. Alternative mechanisms such as *saturation* and *fractional energy loss* are also considered.

We have then discussed whether these CNM effects could explain (or not) the data on J/ψ and ψ' within a coherent picture. We have paid a special attention to the recent PHENIX preliminary data [2] on ψ' production in dAu collisions at $\sqrt{s} = 200$ GeV, which show a strong suppression for central collisions in the mid-rapidity region.

2. Propagation in the cold nuclear matter: J/ψ vs ψ'

The probability for the heavy-quark pair to be broken up in the propagation through the nuclear medium is commonly known as the nuclear absorption. It is usually parametrised by an effective break-up cross section σ_{eff}.

In principle, the larger size of the ψ' compared to the J/ψ tells us that the ψ' should suffer more break-up than the J/ψ. Nevertheless, it is important to remember that the relevant timescale to analyse the pair evolution is its formation time, which, following the uncertainty principle, is related to the time needed – in their rest frame – to distinguish the energy levels of the $1S$ and $2S$ states, $t_f = \frac{2M_{c\bar{c}}}{(M_{2S}^2 - M_{1S}^2)} = 2 \times 3.3$ GeV / 4 GeV$^2 = 0.35$ fm for the ψ. Moreover, t_f has to be considered in the rest frame of the target nucleus, *i.e.* the Au beam at RHIC. The boost factor γ is obtained from the rapidity of the pair corrected by the Au beam rapidity $\gamma = \cosh(y - y_{beam}^{\text{Au}})$ where $y_{beam}^{\text{Au}} = -5.36$ for RHIC.

The formation time for the different rapidities reached at RHIC are given in Table 1. As it

y	$\gamma(y)$	$t_f(y)$	y	$\gamma(y)$	$t_f(y)$
-2.0	14.4	5.1 fm	0.0	106	36.7 fm
-1.5	23.7	8.3 fm	+1.5	476	166 fm
-1.0	39	14 fm	+2.0	786	271 fm

Table 1. Boost and formation-time y-dependence in the Au rest frame of the ψ at $\sqrt{s_{NN}} = 200$ GeV.

is shown there, in the mid and forward rapidity regions at RHIC, t_f is significantly larger than the Au radius. This means that the $c\bar{c}$ is nearly always in a pre-resonant state when traversing the nuclear matter. This implies that the break-up probability should be the same for the ψ' and J/ψ, since these states cannot be distinguished at the time they traverse the nucleus.

3. Gluon shadowing and antishadowing

At high energy, the nucleons in a nucleus *shadow* each other, leading to a modification of the nuclear Parton Distribution Functions (nPDF). This modification, in the form of a shadowing-correction factor, has to be applied to the J/ψ or ψ' yield in dAu collisions. This shadowing factor can be expressed in terms of the ratios R_i^A of the nPDF in a nucleon of a nucleus A to the PDF in the free nucleon.

We shall consider three different shadowing parametrisations: EKS98 [3], EPS08 [4] and nDSg [5] at LO, which have been implemented in our Glauber Monte-Carlo framework JIN [6], already used to describe J/ψ production at RHIC [7, 8, 9]. Our code accounts for improved kinematics corresponding to a $2 \to 2$ ($g + g \to c\bar{c} + g$) partonic process for the ψ production as in the Colour-Singlet Model at LO [10] or the Colour-Octet Mechanism at NLO.

In Fig. 1, the effect of the nPDFs on J/ψ is compared to PHENIX data on J/ψ and ψ' production [2]. We have used the three shadowing parametrisations mentioned above, together with four possibilities for the effective nuclear absortion, namely $\sigma_{\text{eff}} = 0, 2, 4, 6$ mb.

Note that, while the choice of identical break-up cross sections for J/ψ and ψ' is perfectly justified given the argument of large formation times invoked above, the common prejudice that there is a similar shadowing in both cases needs further explanations. It is true that the shadowing corrections, that depends on x and μ_F, should be very similar for both particle production, due to their similar masses. Nevertheless, the difference between their masses –of the order of 0.6 GeV– could induce both a shift on the corresponding x at a given rapidity and on the corresponding μ_F. We have taken the factorisation scale, μ_F, equal to the particle mass, 3.7 GeV for the ψ' and 3.1 GeV for the J/ψ. In Fig. 2, our results for the ratio between the shadowing factors $\frac{R_{\text{dAu}}^{\psi'}}{R_{\text{dAu}}^{J/\psi}}$ for different shadowing parametrisations –EKS98, EPS08, nDSg at LO

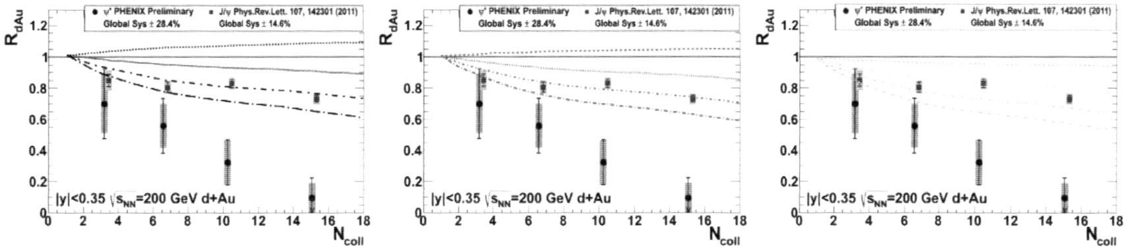

Figure 1. Nuclear modification factor in dAu collisions, $R_{d\mathrm{Au}}$, for J/ψ and ψ' as measured by PHENIX at $\sqrt{s_{NN}} = 200$ GeV versus the number of collisions for the mid-rapidity range $|y| < 0.35$ compared to our computations for four values of the nuclear absorption (from top to bottom: $\sigma_{\mathrm{eff}} = 0, 2, 4, 6$ mb) using EKS98 (left), EPS08 (middle) and nDSg (right).

along with EPS09 at LO and NLO [11] and nDSg at NLO– are shown. Note that the difference

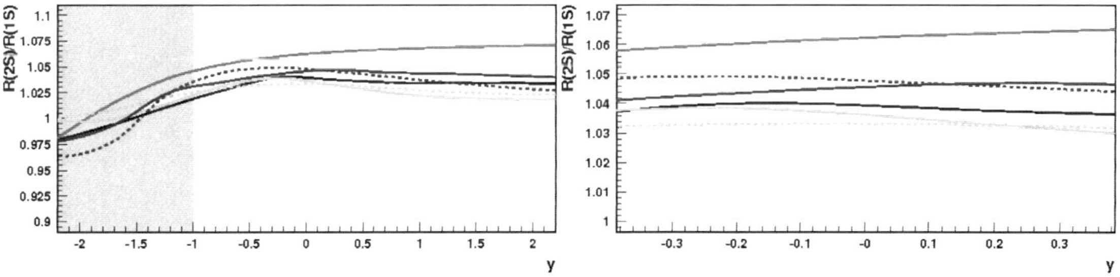

Figure 2. ψ' over J/ψ nuclear modification factors in the rapidity range $-2.2 < y < 2.2$ (left) and in the mid-rapidity region $|y| < 0.35$ (right) using EKS98 (black), EPS08 (red), nDSg (green) and EPS09 (blue) at LO (continous line) and NLO (discontinous line). The shaded zone indicates that t_f is there of the order of the Au radius. In this region, the J/ψ and ψ' absorption may be slightly different.

of the shadowing impact is at most 5 %. Even more important, this difference will favour ψ' production in the mid-rapidity region, in contradiction to what experimental data show.

4. Other effects: gluon saturation and fractional energy loss

The possibility of a gluon saturation as predicted in the framework of the Color Glass Condensate (CGC) [12, 13] is sometimes invoked to have an impact on quarkonium production. We have already discussed [14] that this impact would be negligible in the case of Υ production at RHIC energies. Let us now discuss if it could affect J/ψ and ψ' production. The relevant scale below which one expects significant non-linear effects in the evolution of the gluon distributions in pA is the saturation momentum. It can be evaluated as [15] $Q^2_{s\mathrm{A}} = A^{\frac{1}{3}} \times 0.2 \times \left(\frac{x_0}{x}\right)^{\lambda}$ (in units of GeV^2), with $\lambda \sim 0.2 \div 0.3$ and $x_0 = 0.01$. Typical values for RHIC kinematics are given in Table. 2 together with the ratio of Q_s to the ψ mass. Note that in the backward rapidity region x is always above x_0 –the maximum momentum fraction below which parton saturation can be considered. In the mid and forward rapidity regions, the saturation scale is always below the typical energy scale of the process, namely $m_{\psi'}$ or $m_{J/\psi}$. In particular, in the mid-rapidity region the ratio of Q_s to the ψ mass is $\frac{1}{3}$, which makes unlikely the possibility of specific saturation effects. Only in the most forward rapidity region there is a small probability for phenomena beyond collinear factorisation *i.e.* beyond those encoded in the nPDFs. Moreover, these effects –if they exist– would suppress the J/ψ slightly more than the ψ', in contradiction to the experimental results.

Recently, the possibility of a medium-induced energy loss that scales with the energy E has been pointed out [16]. This *fractional* energy loss would affect the heavy-quark pair produced

y	$Q^{\psi'}_{s\mathrm{Au}}(\mathrm{GeV})$	$\frac{Q^{\psi'}_{s\mathrm{Au}}}{m_{\psi'}}$	$Q^{J/\psi}_{s\mathrm{Au}}(\mathrm{GeV})$	$\frac{Q^{J/\psi}_{s\mathrm{Au}}}{m_{J/\psi}}$
-2.2	< 1	–	< 1	–
-1.2	~ 1	–	~ 1	–
0	$1.0 \div 1.1$	0.3	$1.0 \div 1.1$	0.35
1.2	$1.3 \div 1.4$	$0.35 \div 0.4$	$1.4 \div 1.5$	$0.45 \div 0.5$
2.2	$1.6 \div 1.9$	$0.4 \div 0.5$	$1.7 \div 2.0$	$0.55 \div 0.65$

Table 2. Evaluation of the saturation scale in the Au nucleus, $Q_{s\mathrm{Au}}$, at $\sqrt{s_{NN}} = 200$ GeV in dAu collisions as a function of the ψ' and J/ψ rapidities.

in a coloured state provided that it has long enough time to radiate. The fraction of medium-induced radiated energy is given by $\Delta E/E = \Delta x_1/x_1 \simeq N_c \alpha_s \sqrt{\Delta\langle p_T^2\rangle}/M_T$, where $\Delta\langle p_T^2\rangle$ is the broadening of the radiated gluon from the proton and M_T is the transverse mass of the final-state coloured object. If 100% of the ψ's were produced in a coloured state at RHIC, $\Delta x_1/x_1 \sim 10\%$. Assuming that the corresponding suppression factor is just the ratio between the PDF of the gluon evaluated at its original x'_1, $x'_1 = x_1 + \Delta x_1$ before the loss and the PDF at the resulting x_1: $R_{loss}(x_1, Q^2) = g(x'_1, Q^2)/g(x_1, Q^2)$. Taking modern global-fit gluon PDFs, we find that this *fractional energy loss* could give an extra suppression of the order of 10% in the mid-rapidity region and up to 15% in the forward region. Note that this suppression only applies when the ψ is produced in a coloured state, and that its effect is similar in J/ψ and ψ' production due to their similar masses[1]. Thus, this cannot explain the experimental found either.

In conclusion, we have studied different CNM effects and their impact on J/ψ and ψ' production. We have shown that, at RHIC energies and in the mid-rapidity region, the results from the four CNM effects which we investigated, namely an effective nuclear absorption, the gluon shadowing, the gluon saturation and a fractional energy loss, impact similarly the J/ψ and the ψ', in contradiction to the recent PHENIX preliminary data on ψ' production in dAu collisions at 200 GeV. The abovementioned effects should be taken into account in the interpretation of the upcoming LHC data on J/ψ and ψ' production.

Acknowledgments: This work was partially supported by the Ministerio de Ciencia (Spain) & the IN2P3 (France) (AIC-D-2011-0740).

[1] T. Matsui and H. Satz, Phys. Lett. B **178** (1986) 416.
[2] T. Sakaguchi [PHENIX Collaboration], arXiv:1211.0324 [nucl-ex].
[3] K. J. Eskola, V. J. Kolhinen, C. A. Salgado, Eur. Phys. J. C **9** (1999) 61.
[4] K. J. Eskola, H. Paukkunen, C. A. Salgado, JHEP **0807** (2008) 102.
[5] D. de Florian, R. Sassot, Phys. Rev. D **69** (2004) 074028.
[6] E. G. Ferreiro, F. Fleuret and A. Rakotozafindrabe, Eur. Phys. J. C **61** (2009) 859.
[7] E. G. Ferreiro, F. Fleuret, J. P. Lansberg and A. Rakotozafindrabe, Phys. Lett. B **680**, 50 (2009).
[8] E. G. Ferreiro, F. Fleuret, J. P. Lansberg and A. Rakotozafindrabe, Phys. Rev. C **81** (2010) 064911.
[9] E. G. Ferreiro, F. Fleuret, J. P. Lansberg, N. Matagne and A. Rakotozafindrabe, Few Body Syst. **53** (2012) 27.
[10] S. J. Brodsky, J. P. Lansberg, Phys. Rev. D **81** (2010) 051502.
[11] K. J. Eskola, H. Paukkunen, C. A. Salgado, JHEP **0904** (2009) 065.
[12] E. Iancu, A. Leonidov and L. D. McLerran, Nucl. Phys. A **692** (2001) 583.
[13] E. Ferreiro, E. Iancu, A. Leonidov and L. McLerran, Nucl. Phys. A **703** (2002) 489.
[14] E. G. Ferreiro, F. Fleuret, J. P. Lansberg, N. Matagne and A. Rakotozafindrabe, arXiv:1110.5047 [hep-ph].
[15] J.L. Albacete, N. Armesto, J.G. Milhano, P. Quiroga-Arias and C.A. Salgado, Eur. Phys. J. C **71** (2011) 1705.
[16] F. Arleo, S. Peigne and T. Sami, Phys. Rev. D **83** (2011) 114036.

[1] In fact, the higher mass of the ψ' would lead to a smaller value of $\Delta x_1/x_1$, which is compensated by an increase of the mean value of x_1 when compared to the J/ψ case. This makes the quantitative difference between the J/ψ and the ψ' indistinguishable.

W and Z bosons measurement in PbPb collisions with CMS

Lamia Benhabib
LLR-Ecole Polytechnique
E-mail: `lamia.benhabib@cern.ch`

Abstract. The electroweak bosons W and Z produced in PbPb collisions at LHC at $\sqrt{s}$ = 2.76 TeV have been measured and studied through their leptonic decay channel with the CMS detector. In the following we present the measurement of the Z yields as a function of p_T, rapidity and centrality based on 2011 PbPb data with an integrated luminosity of 150 μb^{-1}, and the W yield as a function of centrality as well as the muon charge asymmetry using 2010 PbPb data corresponding to an integrated luminosity of 7.3 μb^{-1}. The dependence of W and Z production as a function of PbPb collision centrality is shown to scale with the number of NN binary collisions.

1. Introduction

For the first time the LHC offers the new opportunity to study W and Z bosons in heavy ion collisions. From 2010 PbPb data at $\sqrt{s} = 2.76$ TeV corresponding to an integrated luminosity of 7.2 μb^{-1}, the CMS experiment [1] reported that the $Z \rightarrow \mu\mu$ [2] and $W \rightarrow \mu\nu$ [3] do not interact with the hot and dense strongly-interacting medium created in heavy-ion collisions. Indeed, their measurement can allow us to check that their production is unmodified by the medium and thus help us to establish the validity of the Glauber model. Also, they can serve as a reference for other modified probes such as quarkonium production, or the production of an opposite-side jet in Z+jet processes. Finally, their measurement with enough precision can allow to constrain the the initial state of the collision, in particular the nuclear parton distribution functions (PDF).

2. Z bosons

In the Z analysis we require good quality muons with a transverse momentum larger than 20 GeV/c and a pseudo-rapidity $|\eta_\mu| < 2.4$, in order to ensure low background [2]. The dimuon rapidity is limited to $|y| < 2$. Under those conditions, we count 616 Z candidates, i.e. the oppositely charged di-muon pairs which have an invariant mass between 60 to 120 GeV/c^2. The yield of Z bosons (dN_{PbPb}/dy) is then measured as a function of different ranges of Z rapidity (8 bins), p_T (7 bins) and event centrality [1] (6 bins). The centrality classes used here are 50–100%, 40–50%, 30–40%, 20–30%, 10–20% and 0–10% (most central). We calculate the

[1] The centrality of the collisions is determined from the energy deposited in the forward steel/quartz-fibre calorimeters (HF).

yield of $Z \to \mu^+\mu^-$ as a function of p_T ($\frac{d^2N}{dydp_T} = \frac{N(Z\to\mu^+\mu^-)}{\alpha\epsilon N_{MB}\Delta y\Delta p_T}$) and as a function of rapidity ($\frac{dN}{dy} = \frac{N(Z\to\mu^+\mu^-)}{\alpha\epsilon N_{MB}\Delta y}$) where :

- $N(Z \to \mu^+\mu^-)$ (= 616) is the number of counts found in the di-muon invariant mass range of 60–120 GeV/c^2
- α ($\sim$ 70%) and ϵ ($\sim$ 60%) are the acceptance and efficiency corrections
- N_{MB}(= 1.161 billion events) is the number of corresponding minimum bias events corrected for trigger efficiency
- Δy and Δp_T are the bin ranges in consideration

The results are presented in figure 1 (a) and (b) and compared to the pp cross section provided by a next-to-leading-order POWHEG generator, scaled by the proper nuclear overlap function ($d\sigma_{pp}/dy \times T_{\mathrm{AA}}$). To study the Z yield dependence versus centrality we divide N_{MB} by the centrality fraction in consideration. We notice no dependence of the production of Z bosons per binary collision on the number of participants, as shown in 1 (c).

3. W bosons

With 2010 PbPb data at $\sqrt{s} = 2.76$ TeV, the W boson production has been measured through its leptonic muon decay channel ($W^\pm \to \mu^\pm\nu$). The first hint of W observation is the bump present in the muon spectrum, as shown in figure 2 (left). The muon spectrum (solid red dots) is produced by plotting the p_T distribution of all the events selected online with a low (2-3 GeV/c) p_T^μ threshold and $|\eta^\mu| < 2.1$ with strict quality cuts [2] and vetoing on the Z candidate. In this analysis, the energy of the neutrino in the event is estimated by computing a missing transverse momentum $\not{p}_T$ by taking the opposite sign of the vectorial sum of the transverse momenta of all charged particles in the event, with $p_T > 3$ GeV/c. When selecting events with a high p_T muon, the average value of the missing transverse momentum $< \not{p}_T >$ has a very small dependence on centrality and its value is around 40 GeV/c. This value agrees with what is expected for the typical $\not{p}_T$ produced by an undetected neutrino originating from W decay. Therefore, we compute the transverse mass of the W as $m_T = \sqrt{2p_T^\mu \not{p}_T(1-\cos(\phi))}$ where we require $p_T^\mu >$ 25 GeV/c and a $\not{p}_T > 20$ GeV/c and where ϕ is the opening angle between the direction of the high-p_T muon momentum and $\not{p}_T$ vectors. The result is shown in figure 2 (right) where the transverse mass distribution for PbPb data is superimposed on the transverse mass obtained when analyzing pp data with the same procedure (open blue squares). This shows that pp and PbPb results are in a good agreement with a W Pythia signal simulation embedded in PbPb events generated with Hydjet (PYTHIA + HYDJET) (green-hatched histogram) and are almost background free. Residual contamination from other electroweak processes (Z $\to \mu^+\mu^-$ and $W^\pm \to \mu^\pm\nu$) (2%) is subtracted, and the QCD background is estimated to be 1% and included as a systematic uncertainty. We count 275 W^+ and 264 W^- in PbPb and 301 W^+ and 165 W^- in pp by placing a cut of 40 GeV/c^2 on the transverse mass. W, W^+ and W^- yields (dN/dy) in PbPb collisions are scaled by the nuclear overlap function T_{AA} and presented as a function of number of participants (N_{part}) in figure 3 (left). We notice first, that the W production in PbPb collisions does not depend on the centrality of the collision. Second, that the individual W^+ and W^- charge states in PbPb and in pp collisions reflect the different u and d quark content in Pb nuclei and in the proton, what is called the "isospin effect". As a consequence, the nuclear modification factors R_{AA} of W, W^+, W^- are different $R_{AA}(W) = 1.04 \pm 0.07 \pm 0.12$, $R_{AA}(W^+) = 0.82 \pm 0.07 \pm 0.09$, $R_{AA}(W^-) = 1.46 \pm 0.14 \pm 0.16$. The muon charge asymmetry defined as $A = (N_W^+ - N_W^-)/(N_W^+ + N_W^-)$ is studied as a function of the muon pseudo-rapidity for

[2] The quality cut used include the minimum number of hits used in offline reconstruction, the distance in the transverse plane between the muon impact parameter and the primary vertex to be less than 300 μm

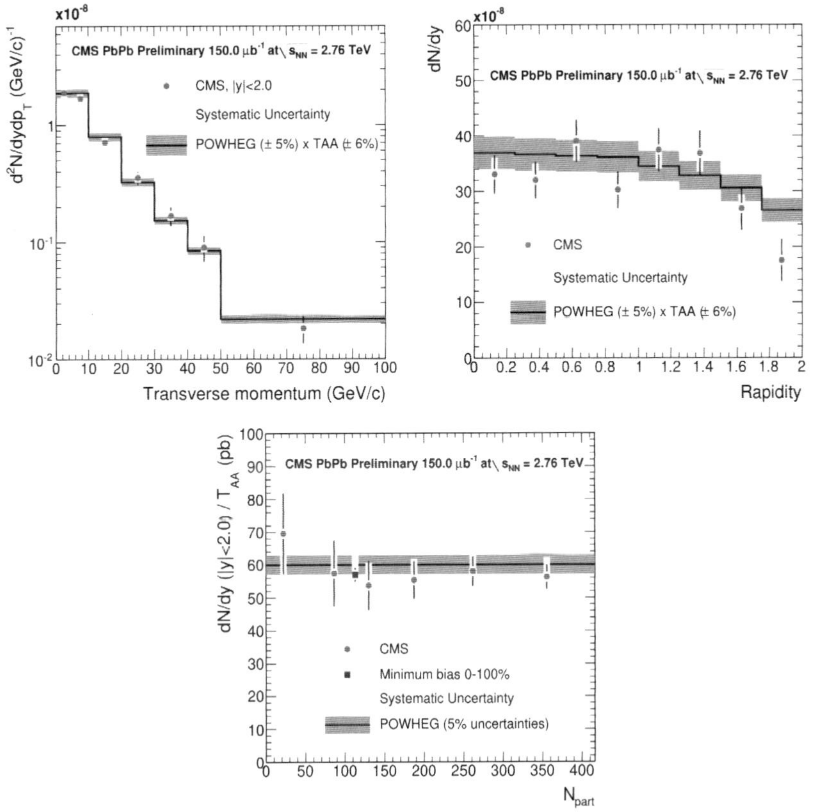

Figure 1. (a) $Z \to \mu^+ \mu^-$ yields versus the Z boson transverse momentum. (b) $Z \to \mu^+ \mu^-$ yields versus the Z boson rapidity. (c) Event centrality dependence of the $Z \to \mu^+ \mu^-$ yield per event divided by the expected average nuclear overlap function T_{AA}. The vertical scale equivalently corresponds to Z yields per binary collision (N_{coll}) times the nucleon-nucleon cross section, and is thus directly comparable to the pp cross section from the POWHEG generator displayed as a black line. On the horizontal axis, event centrality is depicted as the average number of participating nucleons N_{part}. Vertical lines (bands) correspond to statistical (systematic) uncertainties.

both pp and PbPb collisions and shown to agree – within the uncertainties – with the predictions from MCFM generator using MSTW08 PDF for pp and for PbPb data when adding the EPS09 PDF [3] as presented in figure 3 (right).

4. Conclusions

W and Z electroweak boson production is consistent with the binary-collision scaling hypothesis ($R_{AA} = 1$) and can thus serve as reference to modified probes. More statistics could reveal PDF modification. The W from the second run, as well as the electronic decay channels of W and Z are still to be analysed. On the longer term, after the first long LHC shutdown, a significant Z+jet event sample will be available. Even with the isospin effect driving the W^+, W^- to have non-unity values of R_{AA}, the fact that they are flat with centrality demonstrates that the underlying hard scattering processes that lead to W and Z production are well understood thereby adding significance to other CMS observations such as jet suppression.

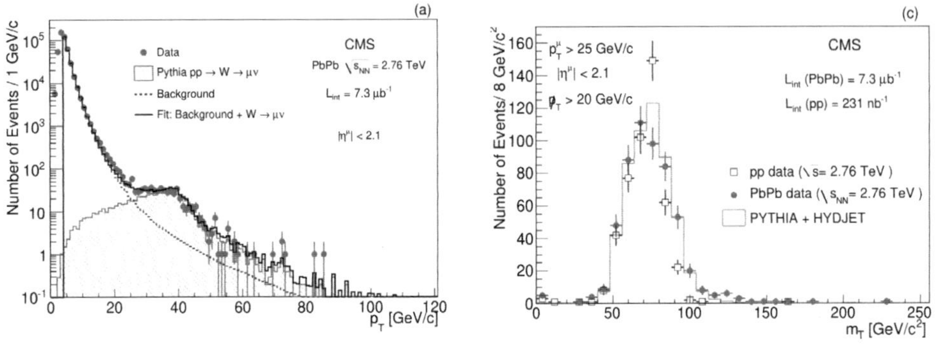

Figure 2. Left: Single-muon transverse-momentum spectrum for $|\mu| < 2.1$ in PbPb data (red points). Signal (green hatched histogram) and background (blue-dashed histogram) contributions are fitted (black-solid line) to the data. Right: Transverse mass distribution for selected events in PbPb (red-filled circles) and pp (blue open squares) data, compared to PYTHIA + HYDJET simulation (hatched histogram). The error bars represent statistical uncertainties .

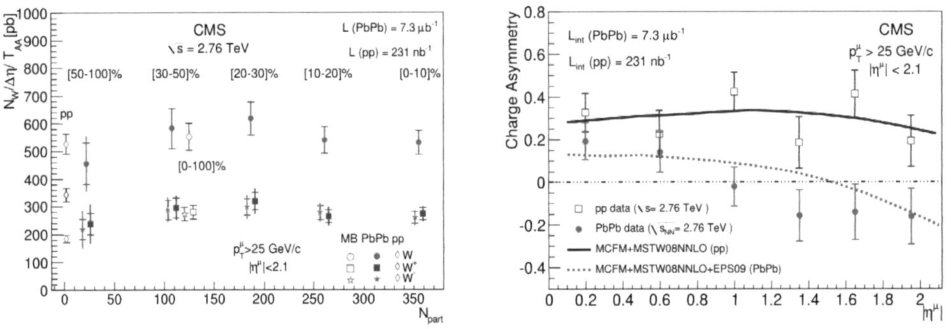

Figure 3. Left: Centrality dependence of normalised W cross sections in PbPb collisions, for all W candidate (red filled points) and separated by charge, W^+ (violet-filled squares) and W^- (green-filled stars). The open symbols at N_{part}= 120 represent the minimum bias events. At N_{part}= 2, the corresponding cross sections are displayed for pp collisions for the same $\sqrt{s}$. The cross sections are given for the phase space region $p_T^\mu > 25$ GeV/c and $|\eta| < 2.1$ [3]. Right: Muon charge asymmetry $A = (N_W^+ - N_W^-)/(N_W^+ + N_W^-)$ as a function of the muon pseudorapidity for PbPb (red points) and pp (blue squares) collisions at 2.76 TeV for muons in the same phase space region.

References

[1] CMS Collaboration, "The CMS experiment at the CERN LHC, JINST 03 (2008) S08004, doi:10.1088/1748-0221/3/08/S08004.

[2] CMS Collaboration, Study of Z boson production in PbPb collisions at nucleon-nucleon centre of mass energy = 2.76 TeV, Phys.Rev.Lett. 106 (2011) 212301, arXiv:1102.5435.

[3] CMS Collaboration, "Study of W boson production in PbPb and pp collisions at $\sqrt{s_{NN}} = 2.76$ TeV", CMS, Phys.Lett.B 715 (2012) 66 doi:10.1016/j.physletb.2012.07.025. arXiv:1205.6334 [nucl-ex]

Journal of Physics: Conference Series **422** (2013) 012020 doi:10.1088/1742-6596/422/1/012020

Thermal and direct photons in PHENIX

Sarah Campbell for the PHENIX Collaboration

Iowa State University, Ames, Iowa, USA

E-mail: campbels@iastate.edu

Abstract. Thermal and direct photons in PHENIX are measured by virtual photons ($\gamma^* \rightarrow e^+e^-$) for p_T 1-5 GeV/c, and real photons for p_T 4-20 GeV/c. In Au+Au, high p_T direct photons show no strong deviation from the T_{AA}-scaled p+p spectrum. The low p_T thermal photon spectra lie above the T_{AA}-scaled p+p fit. Partonic photon production models describe this enhancement with early formation times and high initial temperatures. The Au+Au direct photon elliptic flow, v_2, is large at $p_T < 4$ GeV/c and consistent with zero at $p_T > 4$ GeV/c. Hydrodynamic parton models under-predict the low p_T photon v_2.

1. Introduction

Thermal radiation emitted by the quark-gluon plasma (QGP) provides information on the QGP initial temperature and thermalization process. Thermal photons dominate the direct photon spectrum at low p_T and have positive elliptic flow. Hard processes are the dominant photon source at high p_T. In heavy ion collisions, direct photon yields may be altered by isospin and nuclear parton distribution function effects. There may be additional yields due to jet-medium interactions, and fragmentation photon yields may be suppressed by energy loss of the partons.

2. Prompt photon measurement

The PHENIX detector [1] at RHIC has measured $\sqrt{s}$=200 GeV p+p, d+Au, Cu+Cu and Au+Au collisions. The beam-beam counter (BBC), $3.1 < |\eta| < 3.9$, and the reaction plane detector, $1.0 < |\eta| < 2.8$, separately determine the reaction plane. Two central arm spectrometers extend to ± 0.35 in η and $\pi/2$ in ϕ. Two methods measure photons, real photons in the calorimeters reach high p_T and virtual photons ($\gamma^* \rightarrow e^+e^-$) access lower p_T thermal photons. Real photons shower in the lead-glass Cherenkov and lead-scintillator sampling calorimeters. Electrons are identified by calorimeter showers and rings in the ring imaging Cherenkov detector.

2.1. High p_T direct photons

High p_T direct photons are measured by removing large hadronic decay backgrounds from the inclusive real photons. In the 2006 p+p data, the π^0 tagging method [2] removes these backgrounds, while in the 2004 Au+Au data the double ratio method [3] is used.

The 2006 direct photon p_T spectrum agrees with a NLO pQCD calculation. Using the 2004 data set, the Au+Au high p_T direct photon spectrum shows no apparent deviation from T_{AA}-scaled p+p fit [3], where T_{AA} is the average nuclear thickness function and is calculated using the ratio of the average number of binary collisions in a Glauber Monte Carlo over the total inelastic p+p cross-section of 42 mb. The direct photon nuclear modification factor, R_{AA}, at high p_T is consistent with one for all centralities [3]. Figure 1 shows the high p_T direct photon R_{AA} for 0-5% Au+Au collisions compared with theoretical models that consider initial and final state effects [4][5][6]. The data are consistent with hard scattering processes with initial state effects including isospin effects and nuclear parton distribution functions. High p_T direct photon yields are either unaffected by the final state or a balancing of parton energy loss and medium-induced bremsstralung effects results in no net modification.

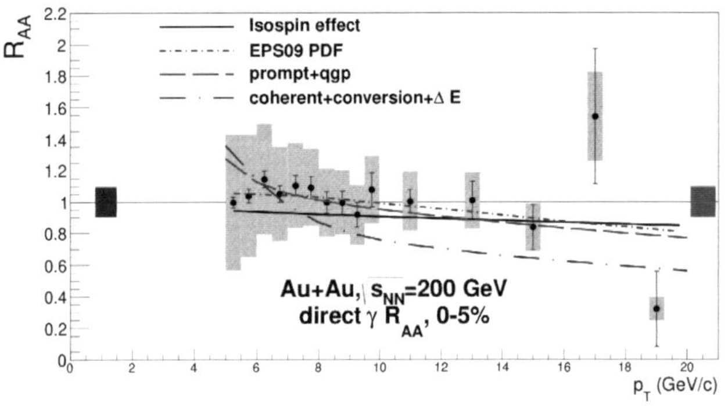

Figure 1. The direct photon R_{AA} for 0-5% Au+Au collisions vs. p_T [3]. Theoretical models consider initial and final state effects [4][5][6].

2.2. Thermal photons from dielectron pairs

Direct photons at lower p_T, where thermal photons dominate, are measured via virtual photons in the dielectron spectrum. These photons have internally converted into electron-positron pairs. When $m \ll p_T$, photon production, which follows the Kroll-Wada formula, is approximately $d^2N_{ee}/dmdp_T \approx 2\alpha/3\pi \times m^{-1}dN_\gamma/dp_T$ [7]. The region $0.1 < m < 0.3$ GeV/c^2 and $p_T > 1$ GeV/c contains virtual photons and excludes $\sim 99\%$ of π^0 backgrounds in minimum bias events.

All electron-positron pairs in an event are made. Combinatorial and correlated backgrounds are generated and statistically subtracted [8]. A cocktail of hadronic backgrounds from π^0, η, ω, η' and ϕ Dalitz decays is simulated with a decay generator assuming m_T scaling of the measured π spectrum and known meson-to-π ratios to set meson yields. The m^{-1} thermal lineshape is also simulated. These simulations are filtered into the PHENIX acceptance.

The direct photon fraction is determined by fitting the cocktail and thermal lineshapes to the mass spectrum in different p_T bins with the function, $(1 - r_\gamma)f_c(m) + r_\gamma f_{th}(m)$, where $f_c(m)$ is the hadronic cocktail, $f_{th}(m)$ is the thermal photon lineshape and r_γ is the only fit parameter, the direct photon fraction [7]. A separate external conversion analysis [9] finds consistent r_γ values. The real direct photon yield is the product of r_γ, the inclusive virtual photon yield, and a real-to-virtual photon production scale factor of $2\alpha/3\pi \times \int m^{-1}dm$ in the 0.1-0.3 GeV/c^2 mass range.

2.3. Spectra

The p+p and d+Au direct photon spectra agree within systematic errors with T_{AA}-scaled NLO pQCD calculations [7][10]. The d+Au nuclear modification factor in d+Au collisions is consistent with one at all p_T, suggesting little to no visible cold nuclear matter effects [10]. In Cu+Cu and Au+Au collisions, a substantial excess in r_γ is seen at low p_T. The low p_T spectrum of direct photons in p+p and Au+Au collisions is shown on the left in Figure 2. The Au+Au spectra are shown with a fit of the p+p spectrum scaled by the average number of binary collisions, $\langle N_{coll}\rangle$, as determined by Glauber Monte Carlo. The inverse slope of the Au+Au low p_T excess provides an average effective temperature of $233 \pm 14 \pm 19$ MeV in minimum bias (MB) events [7]. A recent ALICE result finds an average effective temperature of 304 ± 51 MeV for low p_T

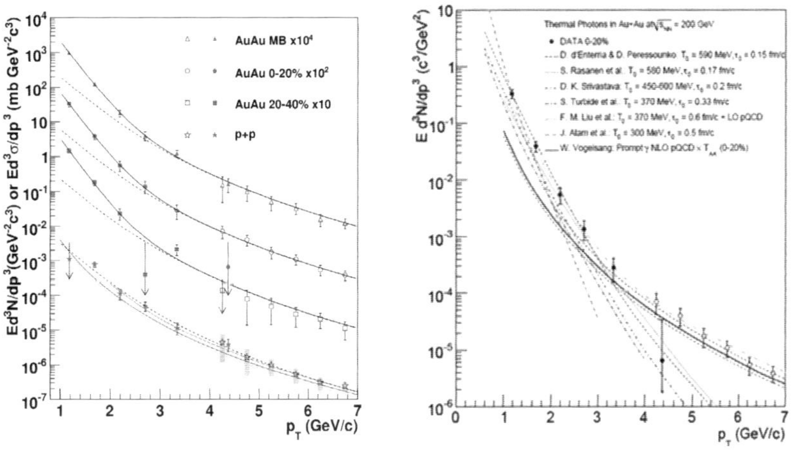

Figure 2. The low p_T direct photon spectrum in p+p and MB, 0-20% and 20-40% Au+Au collisions [7] is on the left. The p+p spectrum is compared to a pQCD calculation [18] and the Au+Au data are shown with a N_{coll}-scaled fit of the p+p spectrum. On the right, the 0-20% Au+Au direct photon spectrum [8] is shown with thermal production models [12]-[17] and a pQCD calculation [18]. Virtual (filled) and real (open) photon measurements are shown.

direct photons in 0-40% $\sqrt{s} = 2.76$ TeV Pb+Pb collisions [11]. Thermal photon emission models [12]-[17], on the right in Figure 2, deduce initial temperatures from 300-600 MeV and formation times of 0.15-0.5 fm/c from the 0-20% Au+Au direct photon spectrum [8].

3. Elliptic flow

The direct photon azimuthal anisotropy provides another metric to study photon production. Elliptic flow, v_2, is the second Fourier component of the direct photon's azimuthal distribution with respect to the event's reaction plane. The thermal photon v_2 is expected to be positive and can provide insight into the viscosity of the QGP [19]. The direct photon elliptic flow, $v_2^{\gamma,dir}$ is determined from the real inclusive photon elliptic flow, $v_2^{\gamma,inc}$, according to Equation 1

$$v_2^{\gamma,dir} = v_2^{\gamma,inc} + \frac{v_2^{\gamma,inc} - v_2^{\gamma,bkg}}{r_\gamma(p_T) - 1} \tag{1}$$

where $v_2^{\gamma,bkg}$ is the elliptic flow of photons from hadron decays and r_γ is the direct photon fraction [20]. The $v_2^{\gamma,bkg}$ is determined by Monte Carlo simulation using the π^0 v_2 and assuming m_T scaling of the hadron p_T spectra and KE_T scaling of the hadron v_2. r_γ is from the real photon method [3] for $p_T > 5$ GeV/c and from the virtual photon method [7] at $p_T < 5$ GeV/c.

Figure 3 shows the direct photon v_2 in MB, 0-20% and 20-40% Au+Au collisions using 2007 data with the BBC reaction plane. At $p_T > 4$ GeV/c, the v_2 is consistent with zero; this agrees with the initial hard scattering picture. Currently, the measured v_2 is unable to resolve the concentration of jet-conversion ($v_2 \approx -0.02$) or fragmentation photons ($0 < v_2 \leq 0.01$) at high p_T. In the thermal photon region, $p_T < 4$ GeV/c, the v_2 is large, comparable to the v_2 of charged hadrons, and decreases with as collisions become more peripheral [20]. The 0-20% Au+Au direct photon v_2 is compared with a hydrodynamic parton model assuming thermalization times of 0.4 (dashed) and 0.6 (dot) fm/c [21]. While the model's shape is similar to the data's, the v_2 magnitude is under-predicted.

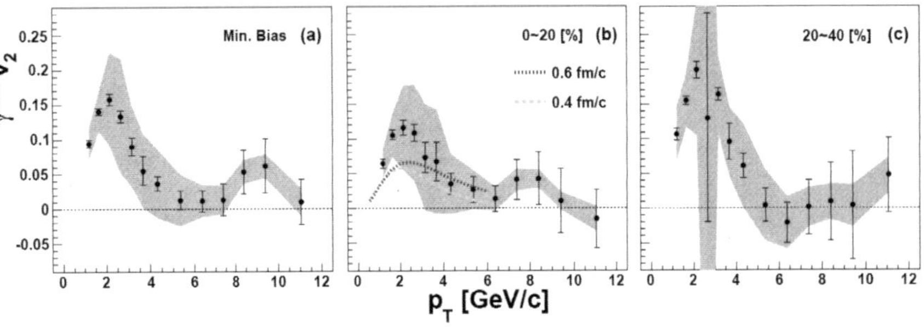

Figure 3. The Au+Au direct photon v_2 vs p_T using the BBC reaction plane [20]. Panel (b) shows a hydrodynamic model with thermalization times of 0.4 (dashed) and 0.6 (dot) fm/c [21].

4. Conclusions

Direct photons at high p_T are from initial hard scattering interactions. In p+p and d+Au, direct photons are well described by NLO pQCD. In Au+Au, the high p_T direct photon spectra undergo no net final state effects and shows zero elliptic flow within systematic errors. The low p_T Au+Au thermal photon spectra show an excess above the T_{AA}-scaled fit to the p+p data. Partonic production models describe the data with a hot initial temperature, 300-600 MeV, and early formation time, 0.15-0.5 fm/c. The thermal photon v_2 is large, comparable to that of charged hadrons. Hydrodynamic parton models are currently unable to reproduce the thermal photon v_2. Later emission may be required for the necessary pressure gradients to build up [22].

Recent PHENIX preliminary 2009 p+p and 2010 Au+Au dielectron spectra [11] using the Hadron Blind Detector may provide improved systematics to measure thermal photons. The lower direct photon-to-π^0 ratio available at RHIC compared to the LHC provides current and future RHIC experiments, such as sPHENIX [23], an advantage in direct photon measurements.

References

[1] K. Adcox, et al. 2003 *Nucl. Instrum. Meth.* **A449** 469
[2] A. Adare, et al. 2012 *Phys. Rev. D* **86** 072008
[3] S. Afanasiev, et al. 2012 *Phys. Rev. Lett.* **109** 152302
[4] S. Turbide, C. Gale, E. Frodermann and U. Heinz 2008 *Phys. Rev. C* **77** 024909
[5] I. Vitev and B.-W. Zhang 2008 *Phys. Lett. B* **669** 337
[6] F. Arleo, K. J. Eskola, H. Paukkunen and C. A. Salgado 2011 *JHEP* **1104** 055
[7] A. Adare, et al. 2010 *Phys. Rev. Lett.* **104** 132301
[8] A. Adare, et al. 2010 *Phys. Rev. C* **81** 034911
[9] R. Petti 2011 *J. Phys.: Conf. Series* **316** 012027
[10] A. Adare, et al. 2012 Direct photon production in d+Au collisions at $\sqrt{s}$=200 GeV arXiv:1208.1234
[11] M. Wilde for the ALICE Collaboration. 2012 Measurement of direct photons in p+p and Pb+Pb collisions with ALICE. submitted for publication in *Nuclear Phys. A* arXiv:1210.5958
[12] D. d'Enterria and D. Peressounko 2006 *Eur. Phys. J* **46** 451
[13] S. Turbide, R. Rapp and C. Gale 2004 *Phys. Rev. C* **69** 014903
[14] P. Houvinen, P. V. Ruuskanen and S. S. Rasanen 2002 *Phys. Lett. B* **535** 109
[15] D. K. Srivastava and B. Sinha 2001 *Phys. Rev. C* **64** 034902
[16] Jan-e Alam, S. Sarkar, T. Hasuda, T. K. Nayak and B. Sinha 2001 *Phys. Rev. C* **63** 021901(R)
[17] F. M. Liu, T. Hirano, K. Werner and Y. Zhu 2009 *Phys. Rev. C* **79** 014905
[18] L. E. Gordon and W. Vogelsang 1993 *Phys. Rev. D* **48** 3136
[19] K. Dusling 2010 *Nucl. Phys. A* **389** 70
[20] A. Adare, et al. 2012 *Phys. Rev. Lett.* **109** 122302
[21] T. Chatterjee and D. K. Srivastava 2007 *Phys. Rev. C* **79** 021901(R)
[22] A. Drees 2012 Hard Probes to be published *Nucl. Phys. A*
[23] A Adare, et al. 2012 sPHENIX: An Upgrade Concept from the PHENIX Collaboration arXiv:1207.6378

Extracting the impact parameter dependence of the nPDFs from the EKS98 and EPS09 global fits

I Helenius[1,2], K J Eskola[1,2], H Honkanen[1,3] and C A Salgado[4,5]
[1]Department of Physics, P.O. Box 35, FI-40014 University of Jyväskylä, Finland
[2]Helsinki Institute of Physics, P.O. Box 64, FIN-00014 University of Helsinki, Finland
[3]The Pennsylvania State University, 104 Davey Lab, University Park, PA 16802, USA
[4]Departamento de Física de Partículas and IGFAE, Universidade de Santiago de Compostela, Galicia-Spain
[5]Physics Department, Theory Unit, CERN, CH-1211 Genève 23, Switzerland

E-mail: ilkka.helenius@jyu.fi

Abstract. As all the globally fitted nuclear PDFs (nPDFs) have been so far impact parameter independent, it has not been possible to calculate the hard process cross sections in different centrality classes consistently with the global analyses. In [1] we have offered a solution to this problem by determining two spatially dependent nPDF sets, `EPS09s` and `EKS98s`, using the A-systematics of the earlier global fits EPS09 and EKS98 and an assumption that the spatial dependence can be written as a power series of the nuclear thickness function. For a data comparison, we have calculated the nuclear modification factor of inclusive neutral pion production in d+Au collisions at RHIC in four centrality bins at midrapidity and compared these to a PHENIX measurement. In addition, we have also performed a similar calculation for inclusive photon production in d+Au collisions at RHIC.

1. Introduction

In the collinear factorization framework [2, 3] the cross section for inclusive k production in a hard process of a heavy ion collision between nuclei A and B can be computed as

$$\mathrm{d}\sigma^{AB\to k+X} = \sum_{i,j,X'} f_i^A(x,Q^2) \otimes f_j^B(x,Q^2) \otimes \mathrm{d}\hat{\sigma}^{ij\to k+X'} + \mathcal{O}(1/Q^2), \tag{1}$$

where f_i^A (f_j^B) is the process-independent nuclear parton distribution function (nPDF) for a parton flavor i (j) in the nucleus A (B) and $\mathrm{d}\hat{\sigma}^{ij\to k+X'}$ represents the partonic pieces which can be computed using perturbative QCD (pQCD). Similarly as the free proton PDFs, also the nPDFs can be determined via a global analysis considering as many different processes as possible. Usually the nPDFs are determined in terms of the free proton PDFs, f_i^N, (x,Q^2) and the nuclear modification of the PDFs, $R_i^A(x,Q^2)$, as

$$f_i^A(x,Q^2) = R_i^A(x,Q^2)\, f_i^N(x,Q^2). \tag{2}$$

So far all the globally analyzed nPDFs have been spatially independent. However, it is reasonable to assume that as the nucleus itself is not spatially uniform, also the nuclear modifications of the PDFs should somehow depend on the position of the nucleon inside the nucleus. In Ref. [1] we have addressed this issue by considering the A-systematics of two globally fitted nPDF sets, EPS09 [4] and EKS98 [5] using the framework discussed in the next section.

2. Framework

First we introduce a nuclear modification of the PDFs, $r_i^A(x, Q^2, \mathbf{s})$, which now depends also on the transverse position $\mathbf{s}$ of the nucleon inside the nucleus. For this we require that the spatial average of the quantity gives back the spatially independent modification,

$$R_i^A(x, Q^2) = \frac{1}{A} \int \mathrm{d}^2\mathbf{s}\, T_A(\mathbf{s})\, r_i^A(x, Q^2, \mathbf{s}), \tag{3}$$

for which we take the values from earlier global fits (EKS98 or EPS09). This, however, does not restrict the form of the spatial dependence in any way, so we have to make an assumption for that. Motivated by the shadowing region at small x, we assume that the $r_i^A(x, Q^2, \mathbf{s})$ can be written as a power series of the nuclear thickness function $T_A(\mathbf{s})$:

$$r_i^A(x, Q^2, \mathbf{s}) = 1 + \sum_{j=1}^{n} c_j^i(x, Q^2)\, [T_A(\mathbf{s})]^j\,, \tag{4}$$

where now the fit parameters $c_j^i(x, Q^2)$ for each parton flavor depend only on x and the scale Q^2 but not on A. This is important for correct mapping of the spatial dependence with the A-dependence of the globally fitted R_i^A. In practice we obtain the values for our fit parameters $c_j^i(x, Q^2)$ by minimizing the χ^2 defined as

$$\chi_i^2(x, Q^2) = \sum_A \left[\frac{R_i^A(x, Q^2) - \frac{1}{A} \int \mathrm{d}^2\mathbf{s}\, T_A(\mathbf{s}) r_i^A(x, Q^2, \mathbf{s})}{W_i^A(x, Q^2)} \right]^2 \tag{5}$$

in a (x, Q^2) grid for all the parton flavors. The weight factor $W_i^A(x, Q^2)$ is set to 1 $(1 - R_i^A(x, Q^2))$ for the EPS09 (EKS98) analysis. For the EPS09 analysis we perform this fitting also for the 30 error sets both in leading (LO) and next-to-leading order (NLO). As can be seen from figure 1, the A-systematics of R_i^A is very well reproduced with the power series ansatz when we consider $A \geq 16$. In our analysis we found out that the first four non-trivial terms of the power series in equation 4 are enough for an accurate fitting in the whole x and Q^2 region, for all the different sets considered.

After the values for the fit parameters $c_j^i(x, Q^2)$ are obtained through this fitting procedure, we can calculate the $r_i^A(x, Q^2, \mathbf{s})$ and determine the new spatially dependent nuclear modification sets, `EPS09s` and `EKS98s`, in which "s" stands for "spatial". These are now published in our website[1]. To illustrate the spatial dependence of the new nPDFs, the gluon modification from EPS09sNLO for lead nucleus is plotted in figure 2 as a function of x and the transverse distance $s = |\mathbf{s}|$. The general feature is that the nuclear effects are larger in the dense center (small s) of the nucleus and smaller at the sparse edge (large s).

3. Applications

The nuclear effects of an observable can be quantified using the nuclear modification factor R_{AA}. The improvement here is that now with the new nPDF sets one can compute the R_{AA} also for different centrality classes in a manner which is consistent with the globally fitted nuclear modifications of the PDFs. Detailed instructions of the implementation of these new nPDFs can be found in Ref. [1].

In figure 3 we have plotted the nuclear modification factor $R_{\mathrm{dAu}}^{\pi^0}$ for inclusive π^0 production in d+Au collisions at $\sqrt{s_{NN}} = 200\,\mathrm{GeV}$ and $y = 0$ as a function of p_T in four centrality classes. The calculations are done in NLO (with the `INCNLO`-package[2]) using the CTEQ6M PDFs [6]

[1] `https://www.jyu.fi/fysiikka/en/research/highenergy/urhic/nPDFs`
[2] `http://lapth.in2p3.fr/PHOX_FAMILY/readme_inc.html`

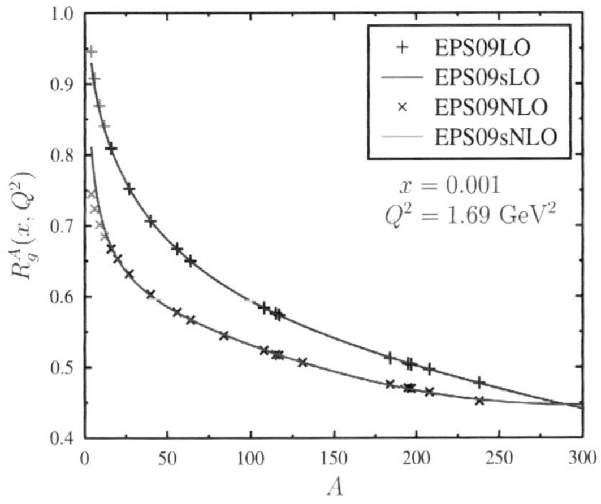

Figure 1. The A-dependence of $R_g^A(x, Q^2)$ at fixed x and Q^2 values from the central sets of EPS09NLO (crosses) and LO (pluses) and from EPS09sNLO (green) and LO (blue). From [1].

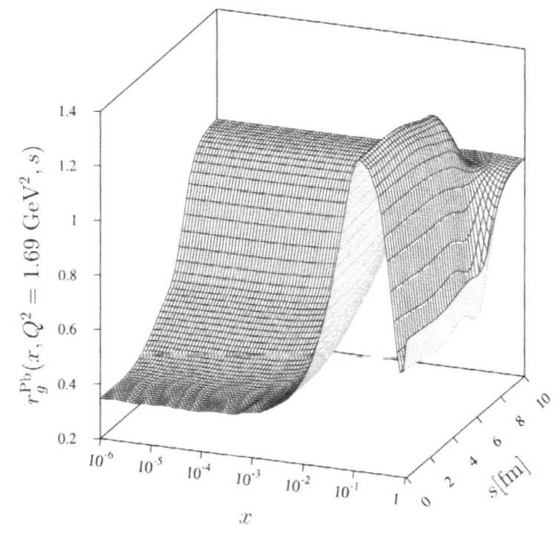

Figure 2. The gluon modification in a lead nucleus $r_g^{\rm Pb}(x, Q^2, s)$ from EPS09sNLO as a function of x and s at the EPS09 initial scale. From [1].

with the EPS09s modifications [1] and three different fragmentation functions (FFs), KKP [7], AKK [8], and fDSS [9]. The uncertainty bands are calculated from the error sets of EPS09s with the fDSS FFs. The centrality classes are defined in terms of impact parameter intervals, which are calculated using the optical Glauber model. The PHENIX datapoints [10] have been scaled by overall factors which all are consistent with the overall normalization uncertainties given by the experiment. The corresponding results for the forthcoming p+Pb collisions at the LHC can be found in [1].

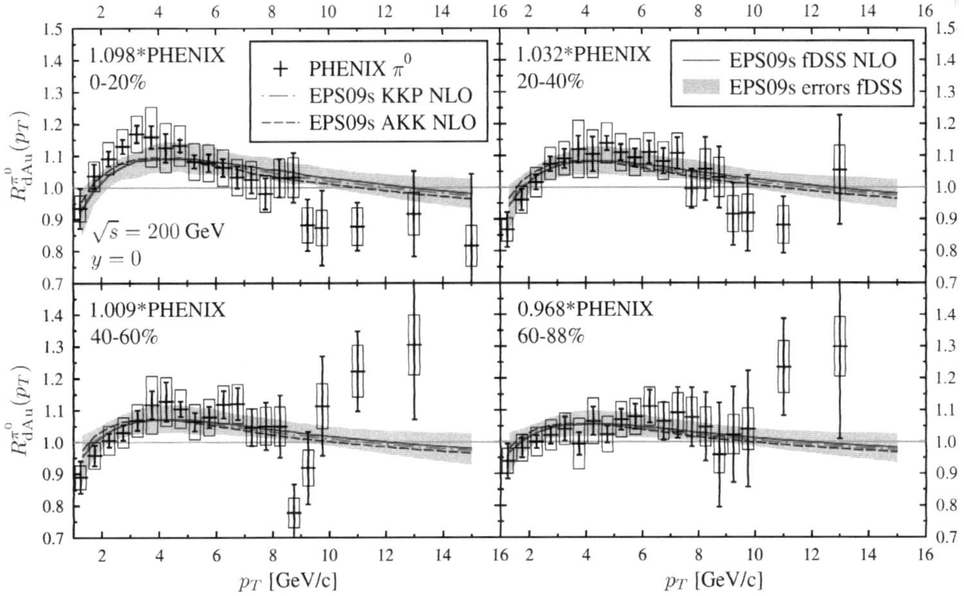

Figure 3. The nuclear modification factor for inclusive π^0 production in d+Au collisions at RHIC in four centrality classes at midrapidity, calculated with the EPS09s nPDFs and different FFs. The data are from PHENIX [10]. From [1].

We have also performed similar calculations for inclusive photon production, $R^{\gamma}_{\rm dAu}$, in d+Au collisions at RHIC, which are shown in figure 4. The results for the minimum bias nuclear modification factor for this process, calculated with different nPDFs, can be found in Ref. [11]. The inclusive photons consists of two components: direct and fragmentation. For the fragmentation component we have used the BFG (set II) FFs [12], otherwise the setup is the same as for the π^0's above. To quantify the isospin effect, we have also plotted the $R^{\gamma}_{\rm dAu}$ using the free nucleon PDFs in each panel. Similarly as for pions, also for the photons we can see that the nuclear modifications are larger for central collisions than for peripheral collisions.

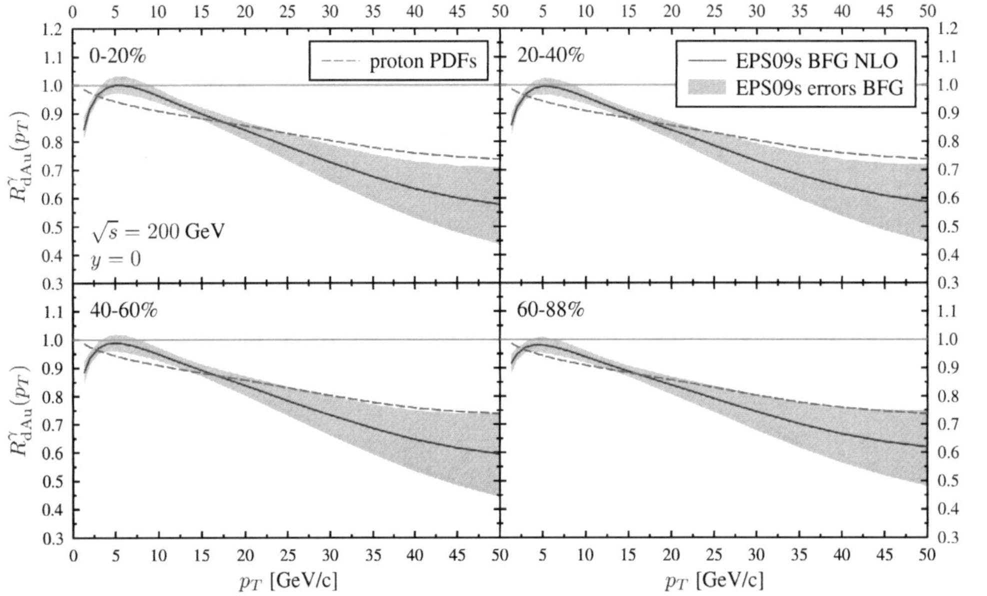

Figure 4. The nuclear modification factor for inclusive γ production in d+Au collisions at RHIC in four centrality classes at midrapidity, computed with nPDFs (EPS09s) and proton PDFs (CTEQ6M). From [13].

Acknowledgments

I.H. and K.J.E. thank the Magnus Ehrnrooth Foundation and Academy of Finland (Project 133005) for financial support. C.A.S. is supported by the European Research Council grant HotLHC ERC- 2001-StG-279579, by Ministerio de Ciencia e Innovación of Spain, and by Xunta de Galicia. H.H. is supported by the U.S. Department of Energy under Grant DE- FG02-93ER40771.

References

[1] Helenius I, Eskola K J, Honkanen H and Salgado C A 2012 *JHEP* **1207** 073 (*Preprint* 1205.5359)
[2] Collins J C, Soper D E and Sterman G F 1988 *Adv.Ser.Direct.High Energy Phys.* **5** 1–91 (*Preprint* hep-ph/0409313)
[3] Brock R *et al.* (CTEQ Collaboration) 1995 *Rev.Mod.Phys.* **67** 157–248
[4] Eskola K J, Paukkunen H and Salgado C A 2009 *JHEP* **04** 065 (*Preprint* 0902.4154)
[5] Eskola K J, Kolhinen V J and Salgado C A 1999 *Eur.Phys.J.* **C9** 61–68 (*Preprint* hep-ph/9807297)
[6] Pumplin J, Stump D R, Huston J, Lai H L, Nadolsky P M *et al.* 2002 *JHEP* **0207** 012 (*Preprint* hep-ph/0201195)
[7] Kniehl B A, Kramer G and Potter B 2000 *Nucl.Phys.* **B582** 514–536 (*Preprint* hep-ph/0010289)
[8] Albino S, Kniehl B A and Kramer G 2008 *Nucl.Phys.* **B803** 42–104 (*Preprint* 0803.2768)
[9] de Florian D, Sassot R and Stratmann M 2007 *Phys.Rev.* **D75** 114010 (*Preprint* hep-ph/0703242)
[10] Adler S S *et al.* (PHENIX Collaboration) 2007 *Phys.Rev.Lett.* **98** 172302 (*Preprint* nucl-ex/0610036)
[11] Arleo F, Eskola K J, Paukkunen H and Salgado C A 2011 *JHEP* **1104** 055 (*Preprint* 1103.1471)
[12] Bourhis L, Fontannaz M and Guillet J 1998 *Eur.Phys.J.* **C2** 529–537 (*Preprint* hep-ph/9704447)
[13] Helenius I and Eskola K J *Work in progress*

Journal of Physics: Conference Series **422** (2013) 012022 doi:10.1088/1742-6596/422/1/012022

Studies on the QCD Phase Diagram at SPS and FAIR

Christoph Blume

Institut für Kernphysik, Goethe-University of Frankfurt,
Max-von-Laue-Str. 1, 60438 Frankfurt am Main, Germany

E-mail: blume@ikf.uni-frankfurt.de

Abstract. A review of results of the energy scan program at the CERN-SPS by the NA49 experiment is given. Presented are observables related to the search for a critical point in the QCD phase diagram and for the onset of deconfinement. Furthermore, the ongoing experimental program of NA61 at the CRRN-SPS and the plans of the CBM experiment at FAIR are discussed.

1. The QCD Phase Diagram

The search for the onset of deconfinement and for a potential critical point in the QCD phase diagram is one of the main subjects of heavy ion physics today. Figure 1 summarizes in a schematic way the features of the QCD phase diagram. While there is a general consensus that the deconfinement phase transition is of the type of a cross over for $\mu_{\mathrm{B}} = 0$ [1], little is known for the case $\mu_{\mathrm{B}} > 0$. Several calculations predict that the cross over line will turn into a first order phase transition [2, 3, 4, 5], thus giving rise to the presence of a critical point in the QCD phase diagram, while other lattice QCD investigations result in a cross over for all μ_{B} [6]. Besides these theoretical considerations there are several experimental programs that try to address this point as well. In order to make progress on this issue, one needs to answer basically three questions:

- Is it possible to locate the onset of deconfinement in the T-μ_{B} plane experimentally?
- Can the data provide any evidence for a first order phase transition at higher μ_{B}?
- Are there observables that would allow to uniquely identify a potential critical point?

By varying the center-of-mass energies $\sqrt{s_{\mathrm{NN}}}$ of the heavy-ion reactions, experiments can scan large regions of the phase diagram. Depending on $\sqrt{s_{\mathrm{NN}}}$ the chemical freeze-out happens at different positions along the the freeze-out curve and therefore the trajectories of the reaction systems in the T-μ_{B} plane have to cross different areas of the phase diagram (as schematically indicated by the arrows in Fig. 1) and might even hit the critical area. In addition, the system size could provide a second control parameter for a systematic scan of the QCD phase diagram. In the following we review the most important experimental facts obtained by the energy scan program of the NA49 experiment at the CERN-SPS and discuss the programs of NA61 at the CERN-SPS and CBM at FAIR.

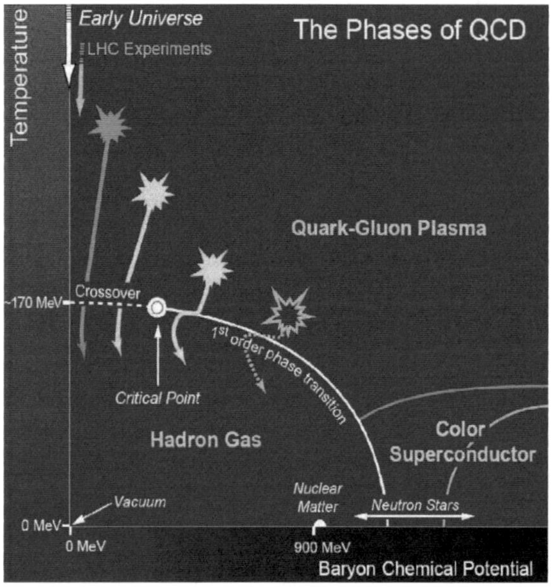

Figure 1. The QCD phase diagram. The arrows roughly indicate the different regions that can be explored by the current and future experimental programs.

2. Result from the SPS Program

In order to locate the onset of deconfinement one needs observables that are sensitive to the change of the number of degrees of freedom that is characterizing the transition from a Hadron Gas (HG) to a Quark-Gluon Plasma (QGP). Promising candidates are e.g. different flow observables. In [7] it has also been argued that the energy dependence of particle yields should reflect a phase transition as well. Indeed, a sharp maximum in the energy dependence of the K^+/π^+ ratio has been found [8], which is not described by any transport model and also not present in p+p data, but might be a signature for a first order phase transition [7].

The evolution of radial flow is to some extent visible the $\sqrt{s_{NN}}$ dependence of the mean transverse mass $\langle m_t \rangle - m_0$, as shown in Fig. 2 for three different particle species. While $\langle m_t \rangle - m_0$ exhibits a steep increase at lower energies, the rise is only very moderate for $\sqrt{s_{NN}} > 7 - 8$ GeV. This sudden change in the energy dependence might be indicative for a change of the equation-of-state [9], even though a further interpretation will require also a more thorough analysis with hydrodynamical models.

When the system trajectories in the T-μ_B plane get close to the critical point critical opalescence should set it, i.e. correlation lengths and susceptibilities will diverge. However, since heavy-ion reactions are naturally limited in their system size, correlations lengths ξ can never exceed the size of the reaction system. Therefore, the critical behavior will be somewhat washed out in comparison to a infinite size system. Nevertheless, the critical behavior might manifest itself in enhanced fluctuations of observable quantities. These could include fluctuations of multiplicity, average p_t, or particle ratios. Of particular interest are fluctuations of conserved quantities, such as strangeness S, baryon number B, and charge Q [11]. A sensitivity to higher orders of ξ can be achieved by measuring higher moments of the related distributions.

Figure 3 shows the energy dependence of multiplicity fluctuations, as measured by the NA49 collaboration. The fluctuations are quantified by the scaled variance $\omega = \text{Var}(n)/\langle n \rangle$, where n is the event-by-event multiplicity. While for all charged particles taken together ω is close to unity, it is slightly below one if negatively and positively charged particles are analyzed

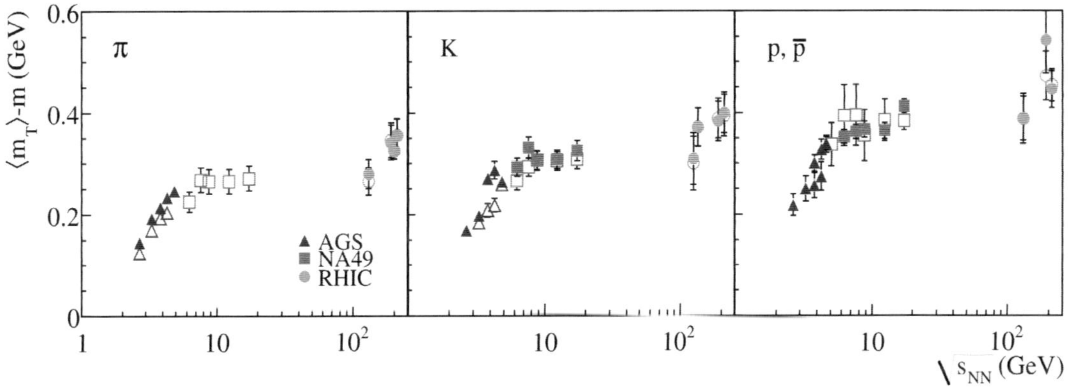

Figure 2. Energy dependence of the mean transverse mass $\langle m_t \rangle - m_0$ measured at midrapidity in central Pb+Pb and Au+Au collisions for $\pi^{\pm}$ (left), $K^{\pm}$ (middle), and p ($\bar{p}$) [8]. Positively (negatively) charged hadrons are indicated by the solid (open) symbols.

separately, i.e. the distributions are a bit narrower than the corresponding Poissonian in these cases. However, in none of the studied charge combinations a significant energy dependence has so far been observed. Also shown in Fig. 3 are theoretical expectations for ω which result from the combination of several assumptions [12]: the position of the critical point in terms of T and μ_B is taken from a lattice QCD calculation [2], the magnitude of the fluctuations at the critical point is based on [14, 15], assuming two different correlation lengths $\xi = 3$ fm (solid line) and $\xi = 6$ fm (dashed line), while the widths of the enhancement around the critical point has been chosen according to [4] as $\sigma(\mu_B) \approx 30$ MeV. Both predictions are at variance with the data, so that up to now there is no evidence for a critical point in fluctuation measurements.

3. The Program of NA61 and CBM

The experimental setup of NA49 was inherited by the NA61 collaboration. After an upgrade of the apparatus (faster readout, addition of a Participant Spectator Detector (PSD)) this collaboration has started to extend the original NA49 program by a two-dimensional, i.e. system size and beam energy, scan. In order to produce beams of light ions the primary Pb ions from the SPS are send onto an internal C-target and the requested nucleus type is selected with high purity from the produced secondary ions by a beam spectrometer. After just finishing the measurement of p+p (six energies), p+C and Be+Be (three energies), the collaboration plans to continue with larger systems (Ar+Ca and Xe+La). First results from the p+p and p+C runs are already published [16, 17].

The study of the QCD phase diagram is also the main pillar of the physics program of the Compressed Baryonic Matter (CBM) experiment at FAIR (Facility for Antiproton and Ion Research). It will be a fixed target experiment at the SIS-100/300 accelerator, covering a beam energy range of 10–45A GeV, and is currently being designed to cope with highest beam luminosities (up to 10^9 ions/s). Apart from the study of all soft hadronic observables, it will therefore also allow to systematically measure various rare probes (e.g. J/ψ, open charm, multi-strange baryons, di-leptons, photons). In a startup phase ($E_{\text{beam}} < 10A$ GeV at the SIS-100) it is planned to move the HADES spectrometer into the CBM cave and to combine it with a first version of the CBM setup. First data taking is planned for 2019. The physics program of CBM is presented in detail in the CBM physics book [18].

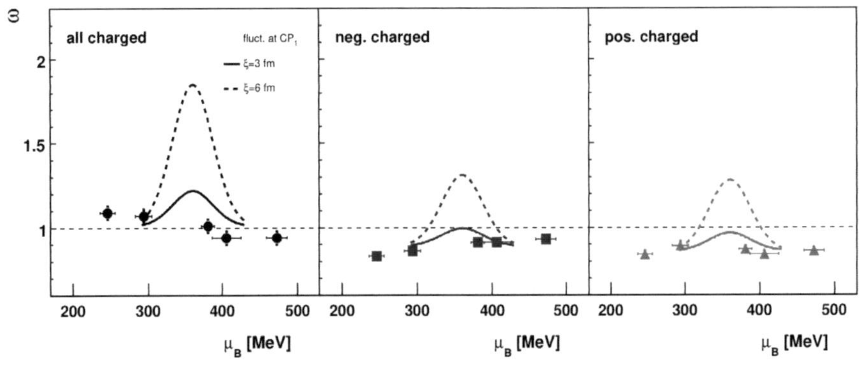

Figure 3. Energy dependence of multiplicity fluctuations, given by the scaled variance ω, for the 1% most central Pb+Pb collisions in the forward rapidity region ($1.1 < y_\pi < y_{\rm beam}$) as measured by the NA49 experiment [10, 12]. The different beam energies are represented by the corresponding $\mu_{\rm B}$ values, determined by a statistical model fit [13]. The lines correspond to predictions for for critical point (see text).

References

[1] Y. Aoki et al., Nature **443**, 675 (2006).
[2] Z. Fodor and S.D. Katz, JHEP **04**, 050 (2004).
[3] C.R. Allton et al., Phys. Rev. **D66**, 074507 (2002).
[4] Y. Hatta and T. Ikeda, Phys. Rev. **D67**, 014028 (2003).
[5] M.A. Stephanov, K. Rajagopal, and E. Shuryak, Phys. Rev. Lett. **81**, 4816 (1998).
[6] P. de Forcrand and O. Philipsen, JHEP **01**, 077 (2007).
[7] M. Gaździcki and M.I. Gorenstein, Acta Phys. Polon. **B30**, 2705 (1999).
[8] C. Alt et al. (NA49 Collaboration), Phys. Rev. **C77**, 024903 (2008).
[9] B. Mohanty et al., Phys. Rev. **C68**, 021901(R) (2003).
[10] C. Alt et al. (NA49 Collaboration), Phys. Rev. **C78**, 034914 (2008).
[11] M.M. Aggarwal et al (STAR Collaboration), Phys. Rev. Lett. **105**, 022302 (2010).
[12] K. Grebieszkow (for the NA49 Collaboration), Nucl. Phys. **A830**, 547C (2009).
[13] F. Becattini, J. Manninen, and M. Gaździcki, Phys. Rev. **C73**, 044905 (2006).
[14] M.A. Stephanov, K. Rajagopal, and E.V. Shuryak, Phys. Rev. **D60**, 114028 (1999).
[15] M.A. Stephanov, *private communication.*
[16] N. Abgrall et al. (NA61 Collaboration), Phys. Rev. **C84**, 034604 (2011).
[17] N. Abgrall et al. (NA61 Collaboration), Phys. Rev. **C85**, 035210 (2012).
[18] B. Friman et al. (CBM Collaboration), Lecture Notes in Physics, **Vol. 814**, 960p (2011).

Journal of Physics: Conference Series **422** (2013) 012023 doi:10.1088/1742-6596/422/1/012023

Physics across the phase diagram of QCD

Sourendu Gupta
Tata Institute of Fundamental Research,
Homi Bhabha Road, Mumbai 400005, India
E-mail: sgupta@tifr.res.in

Abstract. In this talk I present some lattice computations which constrain finite size effects in fireballs resulting from heavy-ion collisions. The results show that early stages of evolution are not constrained by spatial size, whereas freezeout is. The second conclusion is compatible with results of the analysis of fluctuations of conserved quantities, which is a finite-size scaling analysis.

There is incontrovertible evidence now that QCD at zero chemical potential ($\mu = 0$) has no phase transition at finite temperature (T) [1]. However, there are phase transitions in nearby unphysical theories which throw shadows on the thermal behaviour of QCD. One such theory is the version of QCD with massless quarks (the chiral limit) which has a second order phase transition. The other is the version with quark masses all larger than the QCD scale Λ (the quenched theory) which has a first order phase transition. Squeezed between these two platonic worlds, real world QCD shows a gentle crossover between the hadronic world at low T, and the quark gluon plasma at high T: the entropy density in units of temperature changes from about unity at $T = 100$ MeV to about 10 at $T = 200$ MeV [2].

As with any gentle change, if one tries to pin down a point at which the change occurs, then different definitions yield different results. On using a measure of deconfinement (technically, the temperature at which the Polyakov loop susceptibility peaks) one finds $T_c(P) = 170 \pm 4 \pm 2$ MeV. If, on the other hand, one uses a measure of chiral symmetry breaking (technically, the temperature at which the chiral susceptibility peaks) then one finds $T_c(\chi) = 152 \pm 3 \pm 3$ MeV. In both these cases the first error is statistical and the second is an estimate of the error in setting the scale [3].

A direct comparison of lattice predictions for fluctuation variables, which we discuss later, with experimental data finds $T_c(P) = 175^{+1}_{-7}$ MeV [4], in agreement with lattice results of [1]. With this established, it would seem to be useful to explore the physics of this gentle crossover in various ways.

In order to interpret data from heavy-ion collisions one has to make an assumption that the fireball produced in the collisions thermalize. This assumption is tested in various ways. However, it would be good to know whether the fireball is large enough, compared to QCD scales, that thermodynamics is applicable. This involves the measurement of microscopic lengths scales called screening lengths, ξ, in QCD. The physics is intrinsic to any plasma; for any conserved charge, mobile charge carriers in a plasma screen a free charge within a typical scale ξ. As a result, at length scales much larger than ξ, details of interactions can be neglected and thermodynamics is applicable.

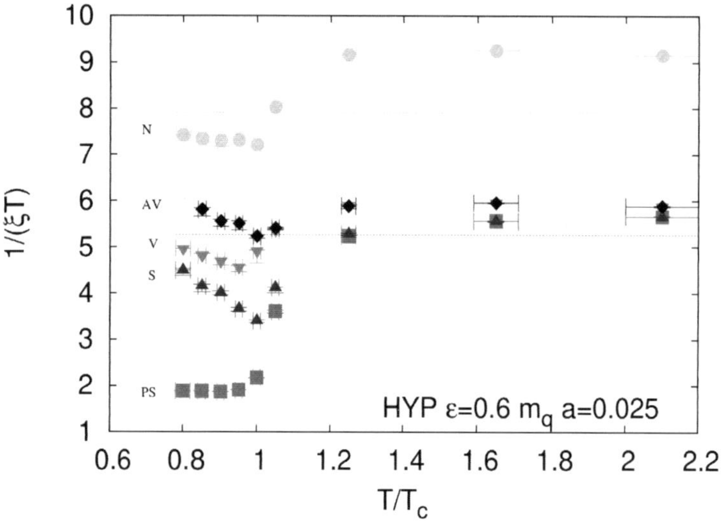

Figure 1. Screening lengths with valence hypercube smeared staggered quarks (HYP) and thin-link staggered sea quarks. Shown are the inverse of the screening lengths in units of the temperature for colour singlet scalar (S), pseudo-scalar (PS), vector (V), axial vector (AV) obtained with a $\overline{q}q$ source and a spin half (N) qqq source. The horizontal lines indicate the values expected in a theory of free quarks, and the scale T_c is $T_c(P)$.

Recently an extended study of screening lengths has been completed in QCD with dynamical staggered quarks with improved scaling properties [5] (see Figure 1) . At temperatures fairly close to $T_c(P)$, for example, at $T = 1.2T_c(P)$, it was observed that all screening lengths are compatible with those in a theory of weakly interacting quarks [6, 7]. As a result at a temperature of about $2T_c(P)$, which could be about the initial temperature of the fireball at large collision energies, all quark correlation lengths are about 0.2 fm. If the fireball is produced in central Pb-Pb collisions, then its size, L, is no less than 6 fm. Since $L \gg \xi$, there is no obstruction to using thermodynamic models for the fireball if the dynamics is fast enough to equidistribute energy among various modes. Collisions at higher energies produce a fireball initially at higher T and closer to $\mu = 0$, so the lattice results indicate that thermodynamics could be applicable at all such beam energies. One may have to re-examine this at lower beam energies, where the initial T is lower and μ is appreciable.

As the fireball expands and cools, one may ask whether finite volume effects become more or less important; the fireball expands, but as it cools the correlation lengths become significantly longer. At the highest RHIC energies freezeout occurs at $T \simeq 165 = 0.94T_c(P)$ MeV, where the smallest $\xi \simeq T$ (see Figure 1). In central collisions, the fireball size at freezeout could be around 10 fm, which means $L/\xi \simeq 8$. As a result, thermodynamics should be good in central collisions, but finite size effects should be important in medium centrality and peripheral collisions.

The study of screening has yielded a little surprise recently [8]. In the quenched theory, where there is a first order deconfining phase transition, no precursors to the transition had been observed. In a study of screening masses in this theory at $T = 0.95T_c(P)$ the screening of various hadrons and their parity partners was examined recently using improved Wilson quarks. In agreement with previous observations [9] it was found that there is no observable thermal effect in the screening of mesons. Extending this measurement to nucleons we again observed no

Journal of Physics: Conference Series **422** (2013) 012023 doi:10.1088/1742-6596/422/1/012023

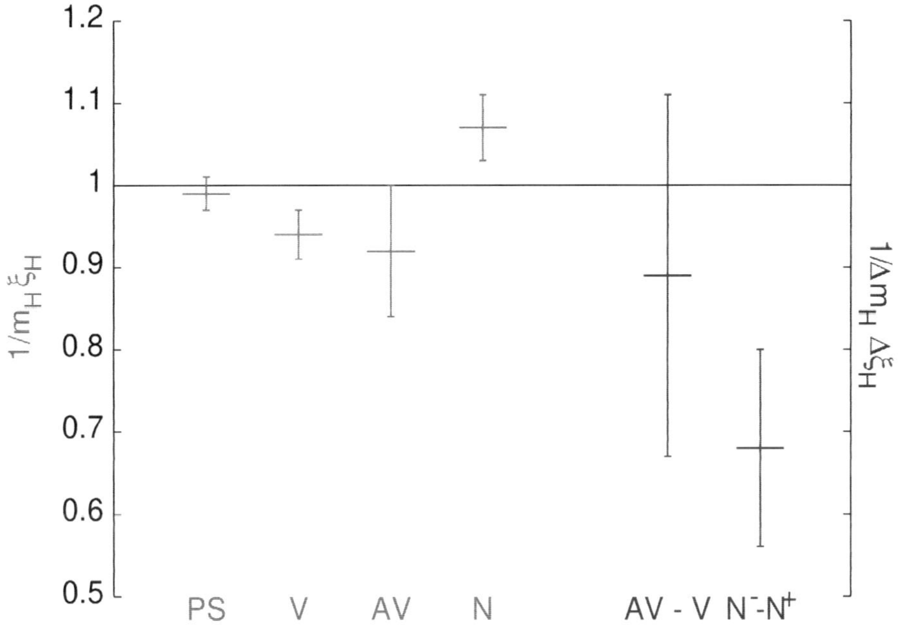

Figure 2. Screening lengths in quenched QCD with clover improved Wilson valence quarks (the horizontal bar denotes the value, and the vertical bar denotes the error). Shown is the inverse of the product of the zero temperature mass and the screening length at $T = 0.95T_c(P)$ in colour singlet pseudo-scalar (PS), vector (V), axial vector (AV) meson channels and in the nucleon (N) channel. Also shown are the splitting of the V and AV and the N and its parity partner N^-, both in units of the zero temperature mass difference between them. The horizontal line shows the values expected if there is no thermal effect.

thermal effects. However, the parity partner of the nucleon seems to be strongly affected by the thermal medium, as a result of which the mass splitting between these two hadrons decreases significantly (see Figure 2). This is something completely new and unexpected, and is worth checking in future.

Information of relevance to experiments is available from lattice computations at finite chemical potential [10]. This comes in the form of a quantitative description of finite size effects. The need for detailed understanding of finite size effects is shown by observations of event-to-event fluctuations of conserved charges.

The argument is simple. It has long been known that thermodynamic fluctuations can be seen only when the volume observed becomes comparable to microscopic dimensions. In the first approximation the spectrum of fluctuations is Gaussian, with computable higher order corrections. The further the deviation is from Gaussian, the more is the importance of microscopic physics. The STAR experiment measured the cumulants of baryon number and found that cumulants up to the 4th order were non-vanishing [11]. In a Gaussian distribution cumulants of order 3 or higher are zero; so finite volume effects studied through these cumulants give information on QCD [12].

Ratios of cumulants which are thermodynamic variables of state were constructed in [12, 13], and computed in lattice QCD in [10]. Agreement of these quantities in QCD and experiment was demonstrated in [11] and this agreement was used to extract $T_c(P)$ in [4].

Ingredients in the lattice computations are the quark number susceptibilities (QNS) [14]

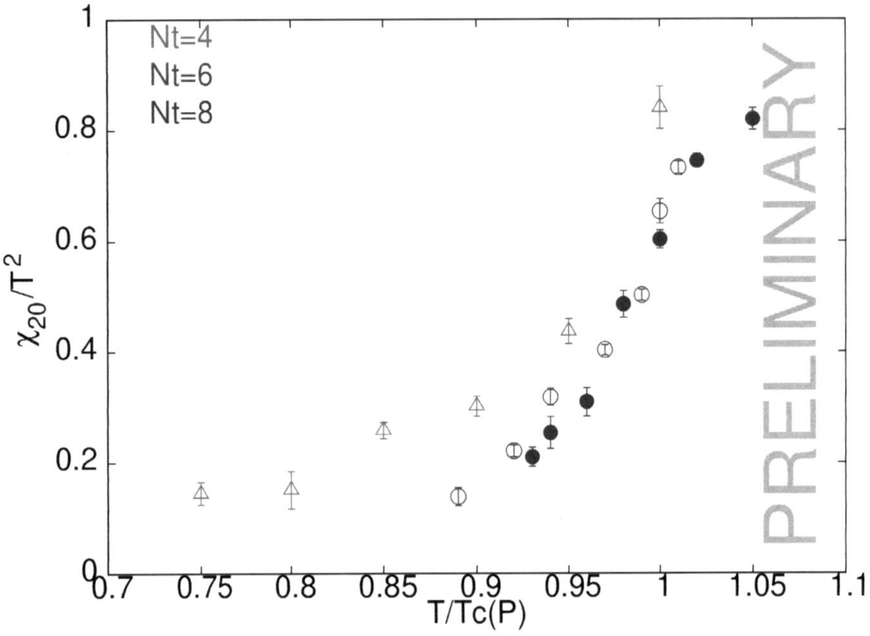

Figure 3. The QNS computed with staggered quarks and lattice spacings $a = 1/(N_tT)$.

and their higher order analogues [15]. In [15, 10] these were computed with a lattice spacing $a = 1/(4T)$ and $a = 1/(6T)$. Now these have been extended to $a = 1/(8T)$ [16] (see Figure 3). Interestingly, the change in the QNS due to this last step is marginal. Computations of the non-linear susceptibilities and further checks are now under way. They will form useful inputs to further analysis of heavy-ion experiments.

References

[1] Aoki Y *et al.* 2006 *Nature* **443** 675
[2] Borsanyi S *et al.* 2010 *J. H. E. P.* **1011** 077
[3] Aoki Y *et al.* 2009 *J. H. E. P.* **0906** 088
[4] Gupta S, Luo X F, Mohanty B, Ritter H G and Xu N 2011 *Science* **332** 1525
[5] Gupta S and Karthik N, in preparation
[6] Eletsky V L and Iofe B L 1988 *Sov. J. Nucl. Phys.* **48** 661
[7] Laine M and Vepsalainen M 2004 *J. H. E. P.* **0402** 004
[8] Padmanath M *et al.*, in preparation
[9] Laermann E, Scheredin S, Wissel S and Karsch F 2007 *Nucl. Phys.* A **785** 261
[10] Gavai R V and Gupta S 2008 *Phys. Rev.* D **78** 114503; *ibid.* 2011 *Phys. Lett.* B **696** 459
[11] Aggarwal M M *et al.* (STAR Collaboration) (2010) *Phys. Rev. Lett.* **105** 022302
[12] Gupta S 2009 *PoS* **CPOD2009** 025
[13] Athanasiou C, Rajagopal K and Stephanov M 2010 *Phys. Rev.* D **82** 074008
[14] Gottlieb S *et al.* 1987 *Phys. Rev. Lett.* **59** 2247
[15] Gavai R V and Gupta S 2003 *Phys. Rev.* D **68** 034506; *ibid.* 2005 *Phys. Rev.* D **71** 114014
[16] Dutta S, Gavai R V , Gupta S, in preparation

Journal of Physics: Conference Series **422** (2013) 012024 doi:10.1088/1742-6596/422/1/012024

Meson Scattering Identifying the BCS-BEC Crossover at Quark Level

Shijun Mao[1] and Pengfei Zhuang[2]
Physics Department, Tsinghua University, Beijing 100084, China
E-mail: maosj08@mails.tsinghua.edu.cn[1]; zhuangpf@mail.tsinghua.edu.cn[2]

Abstract. Instead of the fermion-fermion scattering which identifies the BCS-BEC crossover in cold atom systems, boson-boson scattering is measurable and characterizes the BCS-BEC crossover in QCD superfluid. We study π-π scattering in a pion superfluid described by the Nambu–Jona-Lasinio model. We found that the amplitude of the scattering length drops down monotonously with decreasing isospin density and finally vanishes at the boundary of the phase transition. This indicates clearly a BCS-BEC crossover in the pion superfluid.

There are two kinds of condensed states in usual fermion gas, the Bardeen–Cooper–Shrieffer condensation (BCS) of fermions where the pair size is large and the pairs overlap with each other, and the Bose–Einstein condensation (BEC) of molecules where the pair size is small and the pairs are distinguishable. The BCS wave function can be generalized to arbitrary attraction which leads to a smooth crossover from BCS to BEC [1, 2]. In cold atom systems, the experimental observable to identify the BCS-BEC crossover is the s-wave scattering length between fermions, which is positive in BEC state and negative in BCS state [3, 4, 5, 6].

Recently the study on quantum chromodynamics (QCD) phase structure is extended to finite isospin density. For a QCD system at finite temperature and baryon and isospin density, the phase transitions include not only color deconfinement [7], chiral symmetry restoration [7] and color superconductor [8, 9], but also pion superfluid [10, 11]. The increasing isospin density induces a phase transition from normal nuclear matter to pion superfluid, due to the spontaneous isospin symmetry breaking. By analogy with the usual superfluid, the BCS-BEC crossover in pion superfluid can be theoretically described [12, 13, 14, 15, 16] by the quark chemical potential which is positive in BCS and negative in BEC, the size of the Cooper pair which is large in BCS and small in BEC, and the scaled pion condensate which is small in BCS and large in BEC. However, unlike the fermion-fermion scattering in cold atom systems, quarks are unobservable degrees of freedom, and thus the quark-quark scattering can not be measured or used to experimentally identify the BCS-BEC crossover.

In pion superfluid, the pairs themselves, namely the pion mesons, are observable objects. One can measure the $\pi - \pi$ scattering to probe the properties of the pion condensate and in turn the BCS-BEC crossover. Since pions are Goldstone modes corresponding to the chiral symmetry spontaneous breaking, the $\pi - \pi$ scattering provides a direct way to link chiral theories and experimental data and has been widely studied in many chiral models [17, 18, 19]. Note that pions are also the Goldstone modes of the isospin symmetry spontaneous breaking, the $\pi - \pi$ scattering should be a sensitive signature of the pion superfluid phase transition.

While the perturbative QCD can well describe the properties of the new phases at extremely high temperature and density, the study on the phase transitions at moderate temperature and density depends on lattice QCD calculations [20] and effective models with QCD symmetries. One of the widely used effective models is the Nambu–Jona-Lasinio (NJL) model [21], which is originally inspired by the BCS theory and its version at quark level [22, 23, 24, 25, 26] gives simple and direct description of the dynamic mechanisms of spontaneous chiral symmetry breaking, color symmetry breaking and isospin symmetry breaking. The s-wave $\pi-\pi$ scattering length calculated [17, 18, 19] in the model is consistent with the Weinberg limit [27] and the experimental data [28] in vacuum. In this Letter, we extend the calculation to finite isospin chemical potential and focus on its relation to the BCS-BEC crossover in the pion superfluid.

The Lagrangian density of the two flavor NJL model at quark level is defined as [22, 23, 24, 25, 26]

$$\mathcal{L} = \bar{\psi}\left(i\gamma^\mu\partial_\mu - m_0 + \gamma_0\mu\right)\psi + G\left[\left(\bar{\psi}\psi\right)^2 + \left(\bar{\psi}i\gamma_5\tau_i\psi\right)^2\right] \tag{1}$$

with scalar and pseudoscalar interactions corresponding to σ and π excitations, where m_0 is the current quark mass, G is the four-quark coupling constant with dimension GeV^{-2}, τ_i $(i=1,2,3)$ are the Pauli matrices in flavor space, and $\mu = diag\,(\mu_u, \mu_d) = diag\,(\mu_B/3 + \mu_I/2, \mu_B/3 - \mu_I/2)$ is the quark chemical potential matrix with μ_u and μ_d being the u- and d-quark chemical potentials and μ_B and μ_I the baryon and isospin chemical potentials. At $\mu_I = 0$, the Lagrangian density has the symmetry of $U_B(1)\bigotimes SU_I(2)\bigotimes SU_A(2)$, corresponding to baryon, isospin and chiral symmetry. At $\mu_I \neq 0$, the symmetries $SU_I(2)$ and $SU_A(2)$ are firstly explicitly broken down to $U_I(1)$ and $U_A(1)$, and then the nonzero pion condensate leads to a spontaneous breaking of $U_I(1)$, with pions as the corresponding Goldstone modes. At $\mu_B = 0$, the Fermi surface of $u(d)$ and anti-$d(u)$ quarks coincide and hence the condensate of u and anti-d is favored at $\mu_I > 0$ and the condensate of d and anti-u quarks is favored at $\mu_I < 0$. Finite μ_B provides a mismatch between the two Fermi surfaces and will reduce the pion condensation.

Introducing the chiral and pion condensates $\sigma = \langle\bar{\psi}\psi\rangle$ and $\pi = \langle\bar{\psi}i\gamma_5\tau_1\psi\rangle$ and taking them to be real, the quark propagator $\mathcal{S}$ in mean field approximation can be expressed as a matrix in the flavor space

$$\mathcal{S}^{-1}(p) = \begin{pmatrix} \gamma^\mu p_\mu + \mu_u\gamma_0 - m_q & 2iG\pi\gamma_5 \\ 2iG\pi\gamma_5 & \gamma^\mu p_\mu + \mu_d\gamma_0 - m_q \end{pmatrix} \tag{2}$$

with the dynamical quark mass $m_q = m_0 - 2G\sigma$ generated by the chiral symmetry breaking. By diagonalizing the propagator, the thermodynamic potential can be simply expressed as a condensation part plus a summation part of four quasiparticle contributions [11]. The gap equations to determine the condensates σ (or quark mass m_q) and π can be obtained by the minimum of the thermodynamic potential.

Since the NJL model is non-renormalizable, we should employ a hard three momentum cutoff Λ to regularize the gap equations. In the following numerical calculations, we take the current quark mass $m_0 = 5$ MeV, the coupling constant $G = 4.93$ GeV^{-2} and the cutof $\Lambda = 653$ MeV [29]. This group of parameters ensures the pion mass $m_\pi = 134$ MeV and the pion decay constant $f_\pi = 93$ MeV in the vacuum.

In the NJL model, the meson modes are regarded as quantum fluctuations above the mean field. The two quark scattering via meson exchange can be effectively expressed at quark level in terms of quark bubble summation in RPA [22, 23, 24, 25, 26]. The meson mass $M_i(i = \overline{\sigma}, \overline{\pi}_+, \overline{\pi}_-, \overline{\pi}_0)$ are determined by poles of the meson propagator and their coupling constants $g_{iq\bar{q}}$ are defined as the residues of the propagator at the poles. The Goldstone mode corresponding to the spontaneous broken isospin symmetry is $\overline{\pi}_+$ with zero mass. And the $\overline{\sigma}$ becomes a resonance state in pion superfluid phase, with $g_{\overline{\sigma}q\bar{q}} = 0$ [30].

To the lowest order in $1/N_c$ expansion, where N_c is the number of colors, the invariant amplitude $\mathcal{T}$ is calculated from the diagrams shown in [17, 18, 19]. Different from the

Journal of Physics: Conference Series **422** (2013) 012024 doi:10.1088/1742-6596/422/1/012024

calculation in normal state where both the box and σ-exchange diagrams contribute, the σ-exchange diagrams vanish in the pion superfluid due to the disappearance of the $\overline{\sigma}$ meson. This greatly simplifies the calculation in the pion superfluid.

The scattering amplitude for any channel of the box diagrams can be expressed as

$$i\mathcal{T}_{s,t,u}(k) = -2g^4_{\overline{\pi}q\bar{q}} \int \frac{d^4p}{(2\pi)^4} Tr \prod_{l=1}^{4} [\gamma_5 \tau \mathcal{S}_l] \tag{3}$$

with the quark propagators $\mathcal{S}_1 = \mathcal{S}_3 = \mathcal{S}(p)$, $\mathcal{S}_2 = \mathcal{S}(p+k)$, and $\mathcal{S}_4 = \mathcal{S}(p-k)$ for the s and t channels and $\mathcal{S}_1 = \mathcal{S}_3 = \mathcal{S}(p+k)$ and $\mathcal{S}_2 = \mathcal{S}_4 = \mathcal{S}(p)$ for the u channel. To simplify the numerical calculation, we consider in the following the limit of the scattering at threshold $\sqrt{s} = 2M_{\overline{\pi}}$ and $t = u = 0$, where s, t and u are the Mandelstam variables. In this limit, the amplitude approaches to the scattering length. Note that the threshold condition can be fulfilled by a simple choice of the pion momenta, $k_a = k_b = k_c = k_d = k$ and $k^2 = M^2_{\overline{\pi}} = s/4$, which facilitates a straightforward computation of the diagrams.

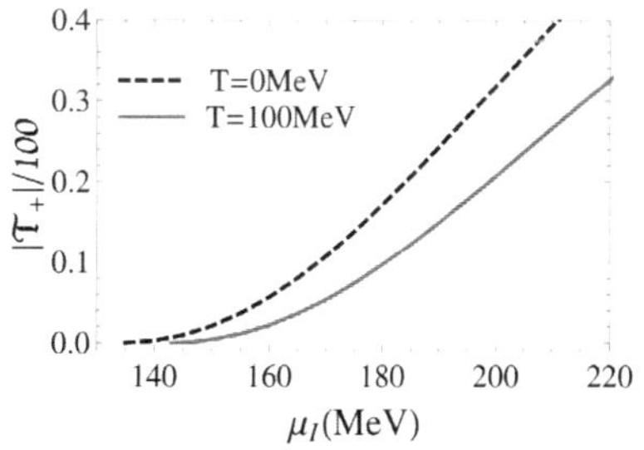

Figure 1. (Color online) The scaled scattering amplitude $\mathcal{T}_+$ as a function of isospin chemical potential μ_I at two values of temperature T.

In Fig.1, we plot the scattering amplitude $|\mathcal{T}_+|$ as a function of isospin chemical potential μ_I at two temperatures $T = 0$ and $T = 100$ MeV, keeping baryon chemical potential $\mu_B = 0$. From (3), the scattering amplitude is governed by the meson coupling constant, $\mathcal{T}_+ \sim g^4_{\overline{\pi}_+ q\bar{q}}$. In the pion superfluid, the meson mode $\overline{\pi}_+$ is always a bound state, its coupling to quarks drops down with decreasing μ_I [30], and therefore the scattering amplitude $|\mathcal{T}_+|$ decreases when the system approaches to the phase transition and reaches zero at the critical value μ_I^c due to $g_{\overline{\pi}_+ q\bar{q}} = 0$ at this point, where the critical isospin chemical potential $\mu_I^c = m_\pi = 134$ MeV at $T = 0$ and 142 MeV at $T = 100$ MeV.

The above μ_I-dependence of the meson-meson scattering amplitude in the pion superfluid with $\mu_I > \mu_I^c$ can be understood well from the point of view of BCS-BEC crossover. We recall that the BCS and BEC states are defined in the sense of the degree of overlapping among the pair wave functions. The large pairs in BCS state overlap each other, and the small pairs in BEC state are individual objects. Therefore, the cross section between two pairs should be large in the BCS state and approach zero in the limit of BEC. From our calculation shown in Fig.1, the $\pi - \pi$ scattering amplitude is a characteristic quantity for the BCS-BEC crossover in pion superfluid. The overlapped quark-antiquark pairs in the BCS state at higher isospin density have large scattering amplitude, while in the BEC state at lower isospin density with separable pairs, the scattering amplitude becomes small. This provides a sensitive observable for the BCS-BEC crossover at quark level, analogous to the fermion scattering in cold atom systems.

In summary, we proposed the meson-meson scattering as an experimentally measurable probe of the BCS-BEC crossover at quark level. Different from the fermion-fermion scattering which is often used to experimentally identify the BCS-BEC crossover in cold atom systems, quark scattering can not be measured and its function to characterize the BCS-BEC crossover in QCD superfluid is replaced by the meson scattering. In the BCS quark superfluid, the large and overlapped pairs lead to large pair-pair cross section, but the small and individual pairs in the BEC superfluid interact weakly. In the frame of a two flavor NJL model at finite temperature and isospin density, we calculated the $\pi - \pi$ scattering length in the pion superfluid. It is large at high isospin chemical potential and drops down monotonously with decreasing isospin chemical potential and finally approaches zero at the border of the pion superfluid, indicating a BCS-BEC crossover.

The meson scattering length $|a_+|$ shown in Figs.1 are obtained in a particular model, the NJL model, which has proven rather reliable in the study on chiral, color and isospin condensates at low temperature. Since there is no confinement in the model, one may ask the question to what degree the conclusions can be trusted. From the general picture for BCS and BEC states, the feature that the meson scattering length approaches to zero in the process of BCS-BEC crossover can be geometrically understood in terms of the degree of overlapping between the two pairs. Therefore, the qualitative conclusion of taking meson scattering as a probe of BCS-BEC crossover at quark level may survive any model dependence. From the recent lattice QCD results [31] at nonzero isospin chemical potential in a canonical approach, the scattering length in the pion superfluid increases with increasing isospin density, which qualitatively supports our conclusion here. We will compute the singular behavior of the higher derivatives of scattering length or the energy density ratio directly, to make further comparison with the lattice results.

Acknowledgement: The work is supported by the NSFC (Grant Nos. 10975084 and 11079024) and RFDP (Grant No.20100002110080).

References

[1] D.M.Eagles, Phys. Rev. **186**, 456(1969).
[2] A.J.Leggett, *Modern Trends in the Theory of Condensed Matter* (Springer-Verlag, Berlin, 1980), p13.
[3] Ph.Courteille, et. al, Phys. Rev. Lett. **81**, 69 (1998).
[4] M.Greiner, C.A.Regal and D.S.Jin, Nature **426**, 537 (2003).
[5] M.W.Zwierlein et. al., Nature **435**, 1047(2005).
[6] T.Bourdel, et. al, Phys. Rev. Lett. **93**, 050401(2004).
[7] For instance, see *Quark-Gluon Plasma*, edited by R.C.Hwa (World Scientific, Singapore, 1990).
[8] M.Alford, K.Rajagopal and F.Wilczek, Phys. Lett. **B422**, 247(1998).
[9] R.Rapp, T.Schaefer, E.V.Shuryak and M.Veldovsky, Phys. Rev. Lett. **81**, 53 (1998).
[10] D.T.Son and M.A.Stephanov, Phys. Rev. Lett. **86**, 592(2001); Phys. Atom. Nucl. **64**, 834(2001).
[11] L.He, M.Jin and P.Zhuang, Phys. Rev. **D71**, 116001(2005).
[12] M.Matsuo, Phys. Rev. C 73 044309 (2006).
[13] J.Margueron, H.Sagawa, and H.Hagino, Phys. Rev. **C76**, 064316(2007).
[14] S.J.Mao, X.G.Huang and P.Zhuang, Phys. Rev. **C79**, 034304(2009).
[15] G.Sun, L.He and P.Zhuang, Phys. Rev. **D 75**, 096004(2007).
[16] C.Mu and P.Zhuang, Phys. Rev. **D79**, 094006(2009).
[17] H.J.Schulze, J. Phys. **G21**, 185(1995).
[18] E.Quack, P.Zhuang, Y.Kalinovsky, S.P.Klevansky and J.Huefner, Phys. Lett. **B34**, 81(1995).
[19] M.Huang, P.Zhuang and W.Q.Chao, Phys. Lett. **B46**, 555(1999).
[20] J.B.Kogut and D.K.Sinclair, Phys. Rev. **D66**, 034505(2002); Phys. Rev. **D70**, 094501(2004).
[21] Y.Nambu and G.Jona-Lasinio, Phys. Rev. **122**, 345(1961); Phys. Rev. **124**, 246(1961).
[22] U.Vogl and W.Weise, Prog. Part. Nucl. Phys. **27**, 195(1991).
[23] S.P.Klevansky, Rev. Mod. Phys. **64**, 649(1992).
[24] M.K.Volkov, Phys. Part. Nucl. **24**, 35(1993).
[25] T.Hatsuda and T.Kunihiro, Phys. Rept. **247**, 338(1994).
[26] M.Buballa, Phys. Rept. **407**, 205(2005).
[27] S. Weinberg, Phys. Lett. 17616 (1966).
[28] D.Pocanic, *Proc. Chiral Dynamics Workshop* in Mainz, Germany, September 1997 and the references therein.
[29] P.Zhuang, J.Hüfner and S.P.Klevansky, Nucl. Phys. **A576**, 525(1994).
[30] X.W.Hao and P.Zhuang, Phys. Lett. **B652**, 275(2007).
[31] W.Detmold, K.Orginos and Z.Shi, Phys. Rev. **D86**, 054507(2012).

Journal of Physics: Conference Series **422** (2013) 012025 doi:10.1088/1742-6596/422/1/012025

QCD critical point, baryon number fluctuations, and final state interactions

Masayuki Asakawa

Department of Physics, Osaka University, Toyonaka, Osaka 560-0043, Japan

E-mail: yuki@phys.sci.osaka-u.ac.jp

Abstract. Baryon number cumulants are important quantities to diagnose the primordial stage of heavy ion collisions. In experiments, however, proton number cumulants have been measured as substitutes. In fact, proton number fluctuations are further modified in the hadron phase and different from those of baryon number. We explore the relation between proton and nucleon number fluctuations in the final state in relativistic heavy ion collisions.

1. Introduction

The order of the phase transition of quantum chromodynamics (QCD) at nonzero temperature (T) is believed to change from crossover [1] to first order as baryon chemical potential ($\mu_{\rm B}$) increases. The existence of the QCD critical point is thus expected in the phase diagram on T-$\mu_{\rm B}$ plane [2]. Experiments to explore the phase structure at nonzero $\mu_{\rm B}$, especially the existence of the critical point, are now ongoing in the energy scan program at the Relativistic Heavy Ion Collider (RHIC) [3, 4, 5], and will also be performed in future facilities. The establishment of the QCD phase structure is an important issue to deepen our knowledge on the matter described by QCD.

Fluctuations, which are experimentally measured by event-by-event analyses in heavy ion collisions, are promising observables to probe properties of created fireballs [6], as their behaviors are sensitive to the state of the matter. Among the fluctuation observables, those of conserved charges can reflect fluctuations produced in earlier stages than non-conserved ones [7, 8, 9]. This is because the variation of a conserved charge in a volume is achieved only through diffusion, which makes the relaxation to equilibrium slower.

The dependences of proton number fluctuations, cumulants up to fourth order, on the beam energy $\sqrt{s}$ are recently measured by STAR collaboration at RHIC [3, 4, 5]. The result looks almost consistent with the prediction of the hadron resonance gas (HRG) model [10]. The proton number, however, is not a conserved quantity, and in fact we will later see that its fluctuations significantly evolve in the hadronic stage, which makes the experimentally-measured fluctuations close to those in the equilibrated hadronic matter. The agreement between the experiments and the HRG model in the proton number fluctuations [5] is in part due to these effects, and hence it does not immediately exclude the slow baryon number diffusion in the hadronic stage. Although the measurement of baryon number, which is a conserved charge, is desirable, its direct experimental measurement has been considered to be impossible because of the difficulty in the detection and identification of neutrons.

In the following, we show that the experimentally-measured proton number fluctuations are nevertheless directly related to baryon number fluctuations in earlier stages, and present concrete formulas that relate baryon number cumulants and these experimental observables. The key observation is that the distributions of (anti-)proton and (anti-)neutron numbers in the final state are well described by binomial distributions. As will be argued in detail later, this observation is justified well at least for RHIC energy, and is expected to hold for $\sqrt{s} \gtrsim 10$GeV. Throughout this article, we use N_X to represent the number of particles X leaving the system after each collision event, where $X = p$, n, and B represent proton, neutron, and baryon, respectively, and their anti-particles, $\bar{p}$, $\bar{n}$, and $\bar{\text{B}}$. The net and total numbers are denoted as $N_X^{(\text{net})} = N_X - N_{\bar{X}}$ and $N_X^{(\text{tot})} = N_X + N_{\bar{X}}$, respectively.

2. Formulation

Let us first briefly consider how the proton number fluctuations evolve in the hadronic stage. The most important process responsible for the variation of the proton number is the charge exchange reactions with thermal pions mediated by $\Delta^+(1232)$ and $\Delta^0(1232)$ resonances, $p(n) + \pi \to \Delta^{+,0} \to n(p) + \pi$. Because of a little energy required and the large cross sections, these reactions proceed even after chemical freeze-out. We note that these reactions do not alter the average abundances, $\langle N_p \rangle$ and $\langle N_{\bar{p}} \rangle$, if the isospin chemical potential vanishes, while they modify the fluctuations of N_p and $N_{\bar{p}}$. Because chemical freeze-out is a concept to describe ratios between particle abundances such as $\langle N_{\bar{p}} \rangle / \langle N_p \rangle$, their existence below the chemical freeze-out temperature, T_{chem}, does not contradict the statistical model. The success of the model, on the other hand, indicates that creations and annihilations of (anti-)nucleons hardly occur below T_{chem}.

The importance of the above reactions below T_{chem} is confirmed by evaluating the mean time of nucleons for these reactions. The mean time is evaluated to be $3-4$fm for $T = 150-170$MeV. One can also check that this mean time hardly changes even for moving protons in the range of momentum $p \lesssim 3T$. On the other hand, dynamical models for RHIC energy predict that protons stay in the hadronic gas and continue to interact for several tens of fm on average at midrapidity [13]. From this result and the lifetime of Δ, it is concluded that nucleons in the fireball indeed undergo this reaction several times on average in the hadronic stage. The ratio of the probabilities that a proton in medium produces a Δ^+ or Δ^0, and then decays into p and n is $5:4$, which is determined by the isospin symmetry. Whereas this probability is not even, after repeating the above processes several times in the hadronic stage nucleons completely forget their initial isospin.

The above observation shows that the evolution of proton number fluctuations in the hadronic stage is dominantly made via the exchanges of the two isospin states of nucleons. Now, we further assert that isospins of all nucleons in the final state are uncorrelated. This statement is well justified when the hadronic medium fulfills the following two conditions: (i) medium effects on the branching ratios and formation rates of Δ are insensitive to the proton and neutron number densities, n_p and n_n (and the same holds for the anti-particle sector as well), and (ii) (anti-)nucleon-(anti-)nucleon interactions generating correlations between two nucleons hardly occur. As we will see later, these two conditions are well satisfied below T_{chem} except for low energy collisions. The probability distribution of finding N_p and N_n ($N_{\bar{p}}$ and $N_{\bar{n}}$) particles in the final state in each event then becomes binomial. Under the isospin symmetry [10], this fact enables to factorize the probability distribution $P(N_p, N_n, N_{\bar{p}}, N_{\bar{n}})$ having N_p, N_n, $N_{\bar{p}}$, and $N_{\bar{n}}$ particles in each event as

$$P(N_p, N_n, N_{\bar{p}}, N_{\bar{n}}) = F(N_{\text{B}}, N_{\bar{\text{B}}}) B(N_p; N_{\text{B}}) B(N_{\bar{p}}; N_{\bar{\text{B}}}), \tag{1}$$

where $B(k; N) = 2^{-N} N!/(k!(N-k)!)$ is the binomial distribution function with an equal probability. On the right hand side (RHS) of Eq. (1) we have used $N_{\rm B}$ and $N_{\bar{\rm B}}$ defined by $N_{\rm B} = N_p + N_n$ and $N_{\bar{\rm B}} = N_{\bar{p}} + N_{\bar{n}}$. For the details of the derivation of Eq. (1), see Ref. [11, 12].

Equation (1) enables to represent baryon number cumulants by those of net and total proton numbers as

$$\langle N_{\rm B}^{(\rm net)} \rangle = 2\langle N_p^{(\rm net)} \rangle, \tag{2}$$

$$\langle (\delta N_{\rm B}^{(\rm net)})^2 \rangle = 4\langle (\delta N_p^{(\rm net)})^2 \rangle - 2\langle N_p^{(\rm tot)} \rangle, \tag{3}$$

$$\langle (\delta N_{\rm B}^{(\rm net)})^3 \rangle = 8\langle (\delta N_p^{(\rm net)})^3 \rangle - 12\langle \delta N_p^{(\rm net)} \delta N_p^{(\rm tot)} \rangle + 6\langle N_p^{(\rm net)} \rangle, \tag{4}$$

$$\begin{aligned} \langle (\delta N_{\rm B}^{(\rm net)})^4 \rangle_c = {} & 16\langle (\delta N_p^{(\rm net)})^4 \rangle_c - 48\langle (\delta N_p^{(\rm net)})^2 \delta N_p^{(\rm tot)} \rangle + 48\langle (\delta N_p^{(\rm net)})^2 \rangle \\ & +12\langle (\delta N_p^{(\rm tot)})^2 \rangle - 26\langle N_p^{(\rm tot)} \rangle, \end{aligned} \tag{5}$$

where $\delta N_X = N_X - \langle N_X \rangle$. Since the RHSs of Eqs. (2) - (5) consist of only $N_p^{(\rm net)}$ and $N_p^{(\rm tot)}$, which are experimentally observable, these are formulas that express baryon number cumulants solely in terms of experimental observables. We remind that no specific form of $F(N_{\rm B}, N_{\bar{\rm B}})$ is assumed in deriving these results.

We remark that $N_{\rm B}^{(\rm net)}$ ($N_{\rm B}^{(\rm tot)}$) in Eqs. (2) - (5) are interpreted to be the sum of all net (total) baryon numbers entering a region in the phase space in the final state of each event. Since the diffusion of baryon number in the hadronic stage is slow [7, 8], the information of primordial fluctuations remains in $F(N_{\rm B}, N_{\bar{\rm B}})$ in Eq. (1) and, as a result, baryon number cumulants.

Next, let us inspect the validity of Eq. (1) in more detail. First, we consider the conditions (i) and (ii) introduced above Eq. (1). At RHIC energy, the Boltzmann approximation is well applied to nucleons below $T_{\rm chem}$ since $T \ll m_N$ and $|\mu_{\rm B}| \ll m_N$. Thus, the Pauli blocking effect can be almost ignored. The Bose factor, on the other hand, has a nonnegligible contribution since $m_\pi \simeq T_{\rm chem}$. The density of pions, however, is more than one order larger than that of nucleons below $T_{\rm chem}$. The Bose factor thus must be insensitive to n_p and n_n, while it leads to the enhancement of the decay of Δ in medium, which acts in favor of the isospin randomization. The large pion density also means that the mean time of a nucleon to form Δ is insensitive to n_p and n_n. The condition (i) is thus well satisfied below $T_{\rm chem}$ at RHIC energy. The validity of the condition (ii) is conjectured from the success of the statistical model as follows. The statistical model indicates that the pair annihilation of a N and an $\bar{\rm N}$ terminates at $T_{\rm chem}$. NN and N$\bar{\rm N}$ reactions are then also expected to terminate there, because the elastic cross section of N$\bar{\rm N}$ is significantly smaller than the inelastic one, and the total cross section of NN behaves similarly as that of N$\bar{\rm N}$ for $E_{\rm cm} < 1$GeV [14]. The condition (ii) thus should also be satisfied for $T < T_{\rm chem}$. Intuitively speaking, in hot medium nucleons are so dilutely distributed that they do not feel other ones' existence, while there are so many pions which can be regarded as the heat bath when the nucleon sector is concerned. The large abundance of pions would also enable to neglect the effects of nucleon isospin fluctuations possibly generated in the early stage. Second, strange baryons, on the other hand, decay via the weak or electromagnetic interaction outside the fireball. In particular, the Λ and Σ are important among them. The Λ decays into p and n with the branching ratio 16 : 9. Provided that the three isospin states of the Σ are produced with an equal probability in medium, the ratio of probabilities that a Σ^+ decays into p and n is about 13 : 12 [14], while a Σ^- decays into $n\pi^-$. A Σ^0 decays into a Λ. If the Λ and Σ multiples are created with an equal probability, the production ration of form their dycays is about 9 : 11. Actually, because of the mass splitting between the Λ and Σ triplets, the production of the Σ triplets are a bit suppressed compared to that of the Λ. This makes the ratio even closer to even. Thus the number of nucleons produced by these dcays can be incorporated into N_p and N_n in Eq. (1). This promotes the nucleon numbers to those of baryons in Eq. (1). Inclusion of higher baryonic resonances and light nuclei such as deuterons will not affect our conclusions owing to their negligible abundances.

Journal of Physics: Conference Series **422** (2013) 012025 doi:10.1088/1742-6596/422/1/012025

While the factorization Eq. (1) is fully established for RHIC energy, the binomiality will eventually break down as the beam energy is decreased. At very low beam energy, pions are not produced enough and nucleons will not undergo charge exchange reactions sufficiently below $T_{\rm chem}$. We deduce that this happens when $T_{\rm chem} \lesssim m_\pi$. When the reactions scarcely occur, the isospin correlations generated at the hadronization will remain until the final state. At low beam energy, also nucleon density becomes comparable with that of pions, and the latter can no longer be regarded as the heat bath to absorb isospin fluctuations of the former. From the $\sqrt{s}$ dependence of the chemical freeze-out line on the T-$\mu_{\rm B}$ plane [15] and considering the validity of these two conditions we deduce that Eq. (1) is well applicable to $\sqrt{s} \gtrsim 10$GeV.

In the argument to derive Eqs. (2) - (5), we have implicitly assumed that the hadronic medium is isospin symmetric. While the effect of nonzero isospin density should be well suppressed for large $\sqrt{s}$ where large number of particles having nonzero isospin charges are produced, at lower energies this effect gives rise to nonnegligible modification to Eqs. (2) - (5). This effect is, as long as the conditions (i) and (ii) introduced above Eq. (1) hold, incorporated to our results by simply replacing the binomial function $B(N_p; N_{\rm B})$ in Eq. (1) with that having probability $k = \langle N_p \rangle / \langle N_p + N_n \rangle$, and similar replacement to $B(N_{\bar{p}}; N_{\bar{\rm B}})$ [12].

3. Conclusion

In the present article, we have derived relations between baryon and proton number cumulants, Eqs. (2) - (5), on the basis of the binomial nature of (anti-)nucleon isospin numbers in the final state. These results enable to immediately determine the baryon number cumulants with experimental results in heavy ion collisions, which will provide significant information on the QCD phase diagram. Though these results are obtained for the isosymmetric case, incorporation of nonzero isospin density is straightforward and is discussed in [12].

This work is supported in part by Grants-in-Aid for Scientific Research by Monbu-Kagakusyo of Japan (No. 23540307).

References

[1] Kanaya K 2010 *PoS LATTICE2010* 012
[2] Stephanov M A *PoS LAT2006* 024
[3] M. M. Aggarwal et al. (STAR Collaboration) 2010 *Phys. Rev. Lett.* **105** 022302
[4] Luo X F *et al.* 2011 *J. Phys. Conf. Ser.* **316** 012003
[5] Mohanty B (2011) *J. Phys.* G **38** 124023
[6] Koch V 2008 Hadronic fluctuations and correlations *Preprint* arXiv:0810.2520
[7] Asakawa M, Heinz U W and Müller B 2000 *Phys. Rev. Lett.* **85** 2072
[8] Jeon S and Koch V 2000 *Phys. Rev. Lett.* **85** 2076
[9] Stephanov M A 2010 *Phys. Rev.* D **81** 054012
[10] Karsch F and Redlich K 2011 *Phys. Lett* B **695** 136
[11] Kitazawa M and Asakawa M 2012 *Phys. Rev.* C **85** 021901
[12] Kitazawa M and Asakawa M 2012 *Phys. Rev.* C **86** 024904
[13] Nonaka C and Bass S A 2007 *Phys. Rev.* C **75** 014902
[14] The Review of Particle Physics, Nakamura K et al. (Particle Data Group) 2010 *J. Phys.* G **37** 075021
[15] Cleymans J and Redlich K 1998 *Phys. Rev. Lett.* **81** 5284

Journal of Physics: Conference Series **422** (2013) 012026 doi:10.1088/1742-6596/422/1/012026

Gluon saturation at higher orders and improvement of kinematics

Guillaume Beuf

Departamento de Física de Partículas and IGFAE, Universidade de Santiago de Compostela, E-15706 Santiago de Compostela, Spain

E-mail: guillaume.beuf@usc.es

Abstract. Standard perturbative calculations lead to pathologically large NLO corrections to low-x_{Bj} evolution equations like BFKL and BK. Using a more refined treatment of kinematics in mixed-space, relevant when gluon saturation sets on, one obtains an improved version of the BK equation, resumming to all orders the most severe of those large higher order corrections.

1. Introduction

Hadronic collisions at very high energy are involving partons with very small momentum fraction in the hadronic wave-functions. Due to the high occupancy of those wee partons (mostly gluons), the phenomenon of gluon saturation occurs: multiple parton scattering is typical, and accompanied by strong color coherence effects. Hence, the collinear factorization (and other standard perturbative QCD formalisms) which involves only one parton from each colliding hadron does not capture the typical physics of collisions in the high-energy limit, with only a semi-hard momentum transfer.

Instead, the coherent multiple scattering effects are taken into account straightforwardly when describing the wee gluons inside each highly boosted hadron as a semi-classical gluon field (see Ref. [1] and references therein). The main formalism based on this idea is the Color-Glass-Condensate effective theory (CGC). Each ultra-relativistic nucleus is described by a random classical shockwave gluon field with a classical statistical distribution, and QCD quantum corrections are resummed within leading logarithmic accuracy (LL) at small momentum fraction by the JIMWLK evolution of the gluon field distribution. The JIMWLK functional equation can also be written as Balitsky's infinite hierarchy of equations.

Inclusive enough observables, like DIS structure functions at low x_{Bj} or single inclusive particle production at forward rapidity in pA collisions, can be expressed in terms of the scattering amplitude of a color dipole on the gluon field of the target. For that object, the Balitsky-Kovchegov (BK) equation [2, 3, 4] gives a safe approximation of the full JIMWLK equation. Adding running coupling effects to the BK equation [5, 6] leads to a successful phenomenological description of DIS data at HERA [7, 8] within the CGC, as well as of forward particle production at RHIC [9].

In the recent years, the calculation of NLO corrections in that framework has been a hot topic. After the running coupling corrections [5, 6], the full set of NLO corrections to the BK equation have been calculated [10]. Later, the calculation of the NLO corrections has been performed for the *impact factor* or *coefficient function* both in the case of DIS structure functions [11, 12] and of forward hadron production in pA collisions [13].

Unfortunately, those NLO calculations cannot be used, in the form in which they are now available, to perform phenomenological studies at full NLO accuracy. Indeed, the BK equation at NLO suffers from the same problem as its linear version, the BFKL equation at NLO [14, 15]: some of the NLO corrections are pathologically large and lead an instability of the solutions. This signals a breakdown of the perturbative expansion as done usually in the Regge limit. Those large NLO corrections are due to the inability of the standard perturbative expansion in the Regge limit to provide results matching smoothly with DGLAP physics in the collinear and in the anticollinear regimes [16]. Hence, the large higher order corrections to the BFKL and BK equations can be resummed to all orders by performing an appropriate matching with the DGLAP equation at LO (or beyond) in the collinear and in the anticollinear regimes [16]. That program has been completed for the BFKL equation both in momentum space [17] and in Mellin space [18]. However, the BK equation is more naturally written in mixed space. The generalization of that resummation to the case of the BK equation requires a significant effort mostly due to the translation to mixed space, and to a lesser extent due to the nonlinearity of the BK equation.

Among the large higher order corrections to be resummed, the ones of purely kinematical origin are the most severe, but also the easiest to deal with [16]. In section 2 of this contribution, the NLO impact factors for DIS [12] are analysed to understand why such kinematical issues arise. And in the section 3, an improved version of the BK equation at LO is proposed, which realizes the resummation of those large kinematical higher order corrections. It corresponds to the mixed space[1] analog of the kinematical constraint [19, 20, 21] in momentum space. It also represents a first step towards a full resummation providing a fully stable and reliable version of the BK equation at NLO.

2. Diagnosing kinematical issues from the explicit NLO impact factors for DIS

The DIS structure functions are linear combinations of the total cross sections for the scattering of a transverse or longitudinal virtual photon off the target, which at strict NLO accuracy in the CGC can be written as [12] (see also [11])

$$\sigma^{\gamma}_{T,L} = 2\,\frac{2N_c\,\alpha_{em}}{(2\pi)^2}\sum_f e_f^2 \int \mathrm{d}^2\mathbf{x}_0 \int \mathrm{d}^2\mathbf{x}_1 \int_0^1 \mathrm{d}z_1 \Bigg\{ \mathcal{I}^{LO}_{T,L}(x_{01},z_1)\,\Big[1-\langle \mathcal{S}_{01}\rangle_0\Big] + \frac{N_c\,\alpha_s}{\pi}\int \frac{\mathrm{d}^2\mathbf{x}_2}{2\pi}\int_0^{1-z_1}\frac{\mathrm{d}z_2}{z_2}\,\mathcal{I}^{NLO}_{T,L}(\mathbf{x}_0,\mathbf{x}_1,\mathbf{x}_2,z_1,z_2)\,\langle \mathcal{S}_{01}-\mathcal{S}_{02}\,\mathcal{S}_{21}\rangle_0 \Bigg\}, \quad (1)$$

where $\mathcal{S}_{ij}$ is the S-matrix for the scattering of a fundamental color dipole with transverse positions $\mathbf{x}_i$ and $\mathbf{x}_j$ off a gluon shockwave, $\langle\ldots\rangle_0$ is the statistical average over the target's gluon field with no LL quantum corrections included, and $x_{ij} = |\mathbf{x}_i - \mathbf{x}_j|$. The LO impact factors $\mathcal{I}^{LO}_{T,L}$ have been known for a long time [22, 23], whereas the NLO ones $\mathcal{I}^{NLO}_{T,L}$ have been calculated in Refs. [11, 12]. The integral over the photon's momentum fraction $z_2 = k_2^+/q^+$ carried by the radiated gluon is logarithmically divergent for $z_2 \to 0$, but in that limit

$$\mathcal{I}^{NLO}_{T,L}(\mathbf{x}_0,\mathbf{x}_1,\mathbf{x}_2,z_1,z_2=0) = \frac{x_{01}^2}{x_{02}^2\,x_{21}^2}\,\mathcal{I}^{LO}_{T,L}(x_{01},z_1)\,. \quad (2)$$

[1] In mixed space, the kinematics of partons is described by their light-cone momentum k^+ and their transverse position $\mathbf{x}$.

Together with an appropriate factorization scheme (including for example a cut-off in k^+), the BK equation

$$\partial_{Y^+} \langle \mathcal{S}_{01} \rangle_{Y^+} = \frac{N_c\, \alpha_s}{\pi} \int \frac{\mathrm{d}^2\mathbf{x}_2}{2\pi}\, \frac{x_{01}^2}{x_{02}^2\, x_{21}^2}\, \langle \mathcal{S}_{02}\mathcal{S}_{21} - \mathcal{S}_{01} \rangle_{Y^+} \tag{3}$$

allows to resum those small z_2 LL contributions. In that case, one should use in the first line of the expression (1) the dipole S-matrix $\langle \mathcal{S}_{01} \rangle_{Y_f^+}$ evolved with the BK equation (3) over a range $Y_f^+ = \log(k_f^+/k_{\min}^+)$. $k_{\min}^+$ is the typical k^+ scale set by the target and k_f^+ an appropriate factorization scale in k^+, such as $k_f^+ = z_1(1-z_1)q^+$.

The LO (resp. NLO) impact factor $\mathcal{I}_{T,L}^{LO}$ (resp. $\mathcal{I}_{T,L}^{NLO}$) contains a factor which suppresses exponentially the large values of $Q^2\, X_2^2$ (resp. $Q^2\, X_3^2$), where

$$X_2^2 = z_1\,(1-z_1)\, x_{01}^2 \quad \text{and} \quad X_3^2 = z_1\,(1-z_1-z_2)\, x_{01}^2 + z_2\,(1-z_1-z_2)\, x_{02}^2 + z_2\, z_1\, x_{21}^2\,. \tag{4}$$

As argued in Ref. [12] the variables $Q^2\, X_2^2$ and $Q^2\, X_3^2$ are the ratios of the formation time of the quark-antiquark or quark-antiquark-gluon Fock components of the photon, resolved by interaction with the target, over the lifetime of the virtual photon. Hence, the interpretation of that exponential suppression is very clear: a Fock state which has not enough time to be formed as fluctuation of the virtual photon within the lifetime of the latter cannot give a non-negligible contribution to the DIS cross sections.

The standard treatment of low z_2 LL with the BK equation discussed previously requires to approximate X_3^2 by X_2^2, in order to obtain the factorization (2) of $\mathcal{I}_{T,L}^{NLO}$. Although exact at $z_2 = 0$, the approximation $X_3^2 \simeq X_2^2$ is not generically correct at small but finite z_2: it is wrong when the gluon is emitted at a so distant transverse position $\mathbf{x}_2$ that $z_1(1-z_1)x_{01}^2 \ll z_2 x_{02}^2 \simeq z_2 x_{12}^2$. In that regime, not only the nice feature of suppression of Fock states too long to form is spoiled by the standard subtraction of LL, but also the term used to subtract the LL contributions from the NLO term in the expression (1) is parametrically larger than both the unsubtracted NLO term and the LO term, which signals a breakdown of this formalism.

Evolution equations like BK and BFKL can be derived from the knowledge of the photon impact factor at arbitrary order but restricted to the case of softer and softer gluons emitted successively [24, 3]. Usually, all the transverse scales are assumed to be of the same order in that context. This assumption (also used in other derivations of those equations) is not completely self-consistent due to the unrestricted integration over transverse momentum or position in the evolution kernel. The resulting issue is essentially the same as found in the study of the NLO photon impact factor: in the parton cascades resummed by the LO BFKL and BK equations, the softer and softer gluons are not always correctly ordered in formation time. The most pathological higher order corrections to the BFKL and BK kernel are then induced by that little inconsistency at LO.

3. Improving the treatment of kinematics in the BK equation

According to our previous discussion, the standard BK equation at LO (3) includes, at large $\mathbf{x}_2$, unphysical contributions from gluons which should not have time to be formed. The first step to cure this problem is to modify the probability density of soft real gluon emission by a color dipole by forbidding emissions with $z_1(1-z_1)x_{01}^2 \ll z_2 x_{02}^2 \simeq z_2 x_{12}^2$. This is the mixed-space analog of the kinematical (*a.k.a.* consistency) constraint of Refs. [19, 20, 21]. In mixed-space, this restriction has been first proposed in Ref. [25], where it was inferred from the structure of n-gluons MHV amplitudes. The resummation scheme proposed here differs however from the one in Ref. [25] in several aspects, most notably in the treatment of virtual corrections.

The choice done in Ref. [26] is to organize the virtual corrections in such a way that the probabilistic interpretation of the dipole cascade [24] is maintained at each order once the resummation is done. Then, the modified virtual corrections can be calculated unambiguously from the modified real gluon emission probability. Performing this task and writing the result as an evolution equation, one gets the improved BK equation with kinematical constraint [26]

$$\partial_{Y^+} \langle \mathcal{S}_{01}\rangle_{Y^+} = \frac{N_c\,\alpha_s}{\pi}\int \frac{\mathrm{d}^2\mathbf{x}_2}{2\pi}\,\frac{x_{01}^2}{x_{02}^2\,x_{21}^2}\,\theta(Y^+ - \Delta_{012})\left\{\left\langle \mathcal{S}_{02}\mathcal{S}_{21} - \frac{1}{N_c^2}\mathcal{S}_{01}\right\rangle_{Y^+-\Delta_{012}} - \left(1-\frac{1}{N_c^2}\right)\langle \mathcal{S}_{01}\rangle_{Y^+}\right\}. \quad (5)$$

A convenient definition for the shift Δ_{012} (but not unique, due to some resummation scheme ambiguity) is

$$\Delta_{012} = \max\left\{0,\, \log\left(\frac{\min(x_{02}^2, x_{21}^2)}{x_{01}^2}\right)\right\}. \quad (6)$$

Due to the theta function in (5), the phase space for gluon emission at large $\mathbf{x}_2$ is severely restricted at small Y^+, but progressively opens up in the course of the Y^+ evolution. The shift of Y^+ in the real term only also contributes to slow down the evolution with respect to the standard BK equation (3).

The largest NLO corrections [10] to the BK equation are indeed resummed into the improved LO equation (5). However, a full resummation of all pathological NLO corrections as in Refs. [17, 18] requires further work. The equation (5) also allows to subtract the LL contributions from the NLO impact factors (1) in a correct way in all the phase-space, by contrast to the equation (3).

Implementing the kinematical improvement (5) together with running coupling corrections [5] should lead to solutions more naturally in good phenomenological agreement with the DIS data, due to a slower Y^+ evolution [7, 8].

References

[1] F. Gelis, E. Iancu, J. Jalilian-Marian and R. Venugopalan, Ann. Rev. Nucl. Part. Sci. **60**, 463-489 (2010).
[2] I. Balitsky, Nucl. Phys. B **463**, 99 (1996).
[3] Y. V. Kovchegov, Phys. Rev. D **60**, 034008 (1999).
[4] Y. V. Kovchegov, Phys. Rev. D **61**, 074018 (2000).
[5] I. Balitsky, Phys. Rev. D **75**, 014001 (2007).
[6] Y. V. Kovchegov and H. Weigert, Nucl. Phys. A **784**, 188 (2007) [hep-ph/0609090].
[7] J. L. Albacete, N. Armesto, J. G. Milhano, P. Quiroga-Arias and C. A. Salgado, Eur. Phys. J. C **71**, 1705 (2011) [arXiv:1012.4408 [hep-ph]].
[8] J. Kuokkanen, K. Rummukainen and H. Weigert, Nucl. Phys. A **875**, 29 (2012) [arXiv:1108.1867 [hep-ph]].
[9] J. L. Albacete and C. Marquet, Phys. Lett. B **687**, 174 (2010) [arXiv:1001.1378 [hep-ph]].
[10] I. Balitsky and G. A. Chirilli, Phys. Rev. D **77**, 014019 (2008).
[11] I. Balitsky and G. A. Chirilli, Phys. Rev. D **83**, 031502 (2011).
[12] G. Beuf, Phys. Rev. D **85**, 034039 (2012) [arXiv:1112.4501 [hep-ph]].
[13] G. A. Chirilli, B. -W. Xiao and F. Yuan, Phys. Rev. D **86**, 054005 (2012) [arXiv:1203.6139 [hep-ph]].
[14] V. S. Fadin and L. N. Lipatov, Phys. Lett. B **429**, 127 (1998).
[15] M. Ciafaloni and G. Camici, Phys. Lett. B **430**, 349 (1998).
[16] G. P. Salam, JHEP **9807**, 019 (1998).
[17] M. Ciafaloni, D. Colferai, G. P. Salam and A. M. Stasto, Phys. Rev. D **68**, 114003 (2003).
[18] G. Altarelli, R. D. Ball and S. Forte, Nucl. Phys. B **742**, 1 (2006).
[19] M. Ciafaloni, Nucl. Phys. B **296**, 49 (1988).
[20] B. Andersson, G. Gustafson and J. Samuelsson, Nucl. Phys. B **467**, 443 (1996).
[21] J. Kwiecinski, A. D. Martin and P. J. Sutton, Z. Phys. C **71**, 585 (1996).
[22] J. D. Bjorken, J. B. Kogut and D. E. Soper, Phys. Rev. D **3**, 1382 (1971).
[23] N. N. Nikolaev and B. G. Zakharov, Z. Phys. C **49**, 607 (1991).
[24] A. H. Mueller, Nucl. Phys. B **415**, 373 (1994).
[25] L. Motyka and A. M. Stasto, Phys. Rev. D **79**, 085016 (2009).
[26] G. Beuf, *in preparation.*

Journal of Physics: Conference Series **422** (2013) 012027 doi:10.1088/1742-6596/422/1/012027

Forward particle production in proton-nucleus collisions at the LHC

H. Fujii

Institute of Physics, University of Tokyo, Komaba 3-8-1, Tokyo, Japan

E-mail: hfujii@phys.c.u-tokyo.ac.jp

Abstract. We present the current theory status on the particle production in the forward region in proton-nucleus (p+A) collisions based on the Color Glass Condensate framework. The cross-section expressed in a factorized form with the dipole amplitude which is related to the unintegrated gluon distribution. Energy dependence of the dipole amplitude is controlled by the nonlinear x-evolution equation and constrained by e+p scattering data at HERA. The transverse geometry of the target is modeled as an assembly of the nucleons distributed with the Woods-Saxon profile. We show our prediction for the nuclear modification factor of the hadron multiplicity in p+A reactions at the LHC energy.

1. Introduction — the CGC picture —

At the Large Hadron Collider (LHC) one can explore hadronic reactions in unprecedentedly wide range of kinematics. Vast majority of particles are produced there with transverse momentum $p_\perp$ at a few GeV/c, which corresponds to particle production from extremely small Bjorken's $x = p^+/P^+$ components[1] because $P^+ \propto \sqrt{s}$. The small-x partons emerge from the short-time quantum fluctuations in light-cone perturbation theory. The higher the collision energy is, the shorter-lived fluctuations (partons) participate in the reaction according to the Lorentz time dilation (the BFKL cascade). Such a growth of the *gluon* distribution at small x is observed in e+p scattering experiments at HERA. Moreover, at some point in the density growth, parton merging should become operative to make the dynamics nonlinear even in the small coupling framework.

The QCD effective theory for the dense small-x degrees is formulated as Color Glass Condensate (CGC) [1], which describes a new universal aspect of QCD – emergence of a dynamic scale for the parton density per unit transverse area, so-called *saturation scale*, $Q_s^2(x)$. In e+p scatterings, the quark pair dissolved from the virtual photon, interacts coherently with the color field created by this dense gluon system of density $Q_s^2(x)$. This CGC picture involves

- multiple scatterings in the target (Cronin peak),
- gluon density depletion compared to the linear BFKL cascade.

In case of a nucleus of mass number A, the pair probes the color charge in the nucleons in the target as a whole. Dense valence charge desnity $xG(x,Q^2)/\pi R^2 \propto A^{1/3}$ enhances the nonlinearity of the dynamics, yielding stronger gluon depletion in the nuclear target even at moderate values of x.

[1] Here $p^\pm$ are the light-cone momenta of the parton, $p^\pm = (p^0 \pm p^3)/\sqrt{2}$, and $P^\pm$ of the incident nucleon.

Journal of Physics: Conference Series **422** (2013) 012027 doi:10.1088/1742-6596/422/1/012027

The significance of p+Pb collisions at the LHC is:

- the control experiment to A+A collisions for disentangling the initial from final state effects
- to check the relevance of the CGC by comparing p+A with p+p in wider kinematics.

2. CGC phenomenology

A current practical tool for the CGC analysis is the Balitsky-Kovchegov equation improved with running coupling corrections (rcBK equation) [2, 3]. The rcBK equation describes the x-evolution of the dipole amplitude $\widetilde{\mathcal{N}}_F(k,x)$ in the fundamental representation, resumming the $\alpha_s \ln(1/x)$ corrections and including the gluon recombination effects. The running coupling improvement gives rise to x-dependence of $Q_s^2(x)$ consistent with the empirical estimate[4, 5].

2.1. Constraining parameters

The amplitude $\widetilde{\mathcal{N}}_F(k,x)$ is confronted with the compiled e+p data at HERA within the dipole model in Ref. [6]. Setting the initial condition

$$1 - \mathcal{N}(r, x_0 = 0.01) = \exp\left[-\frac{(r^2 Q_{s0}^2)^\gamma}{4} \ln\left(\frac{1}{r\Lambda_{\rm QCD}} + e\right)\right] \tag{1}$$

with the transverse size r, they successfully fit the data with certain parameter sets. Here we use the modified sets as given in Table 1 which are obtained with a smooth regulator for the coupling constant, $\alpha_s(r) = 4\pi/(9\ln[4C^2/(r^2\Lambda_{\rm QCD}^2) + a])$ (constant a is adjusted by $\alpha_s(\infty) = \alpha_{\rm fr}$). The result is insensitive to the infrared region as far as $Q_s^2 \gg \Lambda_{\rm QCD}^2$. The parameter $\gamma > 1$ is needed to fit the $p_\perp$ slope (Fig. 1).

We still need a model for a nuclear target. Following Ref. [7], we distribute the nucleons stochastically in the nucleus with the Woods-Saxon profile, and set locally the saturation scale as

$$Q_{s0,A}^2 = N \times Q_{s0}^2 \tag{2}$$

with N the number of the nucleons along the trajectory of the projectile nucleon (Fig. 2).

set	Q_{s0A}^2/GeV^2	γ	$\alpha_{\rm fr}$	C
MV	0.2	1	0.5	1.0
h	0.1597	1.118	1.0	2.47
h'	0.157	1.101	0.8	1.0

Table 1. Fit parameters. Set "MV" is for comparison. $\Lambda_{\rm QCD} = 0.241$ GeV.

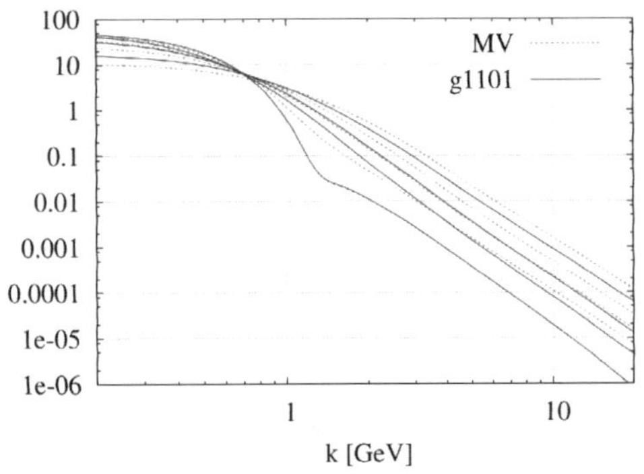

Figure 1. Tail of proton's $\widetilde{\mathcal{N}}_F(k,x)$ at $\ln(x_0/x) = 0, 1.5, 3, 6$ for sets h' and MV.

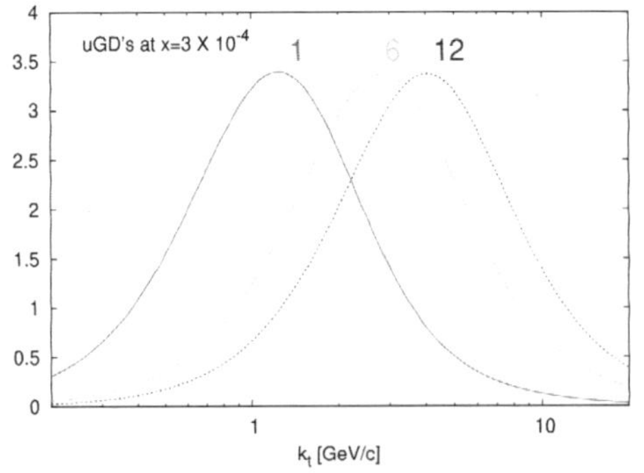

Figure 2. $k^2\widetilde{\mathcal{N}}_A(k,x)$ at $x = 3\times10^{-4}$ evolved from $x_0 = 0.01$ with $Q_{s0,A}^2 = (1, 6, 12) \times Q_{s0}^2$.

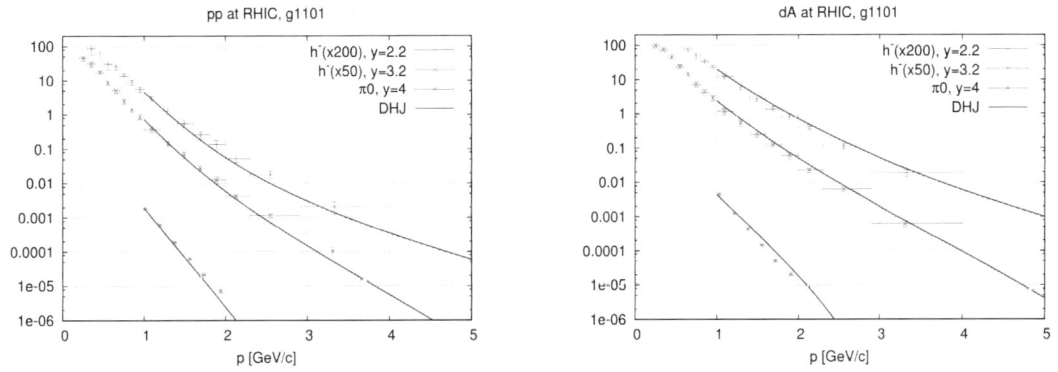

Figure 3. Hadron production in p+p (left) and d+Au (right) reactions at RHIC [14, 15].

2.2. Forward particle production formula

The sets in Table 1 are examined with the LHC p+p data in the mid-rapidity region, using the $k_\perp$ factorization formula. The sets with $\gamma \sim 1.1$ result in a good description for the data[8, 9].

The forward particle production with rapidity $y > 0$, where $x_{1,2} = (p_\perp / z\sqrt{s})e^{\pm y}$ in $2 \to 1$ kinematics, has more sensitivity to the small-x_2 distribution. The cross-section is given in the factorized formula of Dumitru, Hayashigaki, and Jalilian-Marian (DHJ)[10]:

$$\frac{dN}{dy_h d^2 p_\perp} = \frac{K}{(2\pi)^2} \sum_{i=q,g} \int_{x_F}^{1} \frac{dz}{z^2}\, x_1 f_{i/p}(x_1, p_\perp^2)\, \widetilde{\mathcal{N}}_i\left(\frac{p_\perp}{z}, x_2\right) D_{h/i}(z, p_\perp^2)\,, \tag{3}$$

where y_h and $p_\perp$ are the rapidity and transverse momentum of the produced hadron h, respectively. The $f_{i/p}(x_1, Q^2)$ is the collinear distribution function for the large-x_1 parton i, $\widetilde{\mathcal{N}}_{F,A}(k, x_2)$ the fundamental/adjoint dipole amplitude at small x_2, and $D_{h/i}$ the fragmentation function of a parton i into a hadron h with the momentum fraction z. This DHJ formula is obtained in the LO accuracy, and the constant parameter K may absorb some part of higher order effects. We shall keep the parameters x_0 and K be the same for p and A, although we need K dependent on the produced hadron species. Once we fix these in e+p and p+p collisions and $Q^2_{s0,A}$ by Eq. (2), we have no additional parameters.

3. Results

We use the parametrizations of CTEQ6M NLO[11] and DSS NLO[12] for $f_{i/p}$ and $D_{h/i}$, respectively, and choose the factorization scale to $\mu^2 = p_\perp^2$. We show in Fig. 3 our results on transverse momentum distributions of negatively charged hadrons h^- at pseudo-rapidities $\eta = 2.2$ and 3.2, and of neutral pions π^0 at $\eta = 4$ in pp and d-Au collisions at $\sqrt{s} = 200$ GeV[13, 9]. Data at $\eta = 2.2$ and 3.2 (4) are taken from BRAHMS[14] (STAR[15]). The set h' with $K = 1.0$ (0.4) describes the forward particle multiplicities of $h^-(\pi^0)$ in p+p and d+Au collisions at the same time without changing any parameters in the model.

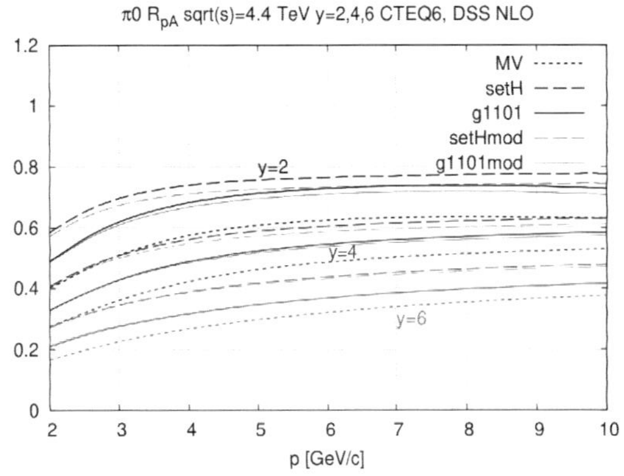

Figure 4. Nuclear modification factor $R_{\rm pPb}$ at $y = 2, 4, 6$ at the LHC ($\sqrt{s} = 4.4$TeV) with several parameter sets.

The nuclear modification factor is defined as the ratio of the cross-sections in p(d)+A and p+p collisions normalized by the number of nucleon collisions; $R_{\mathrm{p(d)}A} = d\sigma_{\mathrm{p(d)}A}/(\langle N_{\mathrm{coll}}\rangle d\sigma_{\mathrm{pp}})$. The forward suppression of R_{dAu} was reported at RHIC in Refs. [14, 15]. Our calculation based on the CGC qualitatively reproduces this suppression at RHIC as a saturation effect [16, 13]. However, it is somewhat sensitive to the initial setup at $x = x_0$ because a subtle balance is needed between the Cronin-like enhancement due to multiple scattering and the saturation due to parton merging within the narrow evolution interval. At the LHC energy, one can expect a much wider evolution interval; x_2 can be as small as $x_2 \sim 10^{-6}$ at $y = 6$. In Fig. 4 we show our present expectation for $R_{\mathrm{pPb}}(p_\perp)$ at $y = 2$, 4 and 6 at $\sqrt{s} = 4.4$ TeV. Despite ambiguities coming from the parameter uncertainty, we see that the saturation effect leads to systematically stronger suppression of $R_{\mathrm{pPb}}(p_\perp)$ in forward rapidity region. This suppression is quantitatively different from the estimate based on nuclear parton distributions (see Ref. [9] for more discussions).

4. Summary

We have evaluated the $p_\perp$ spectrum of the single hadron production in the forward region at RHIC and the LHC energies, using the gluon distributions obtained by the rcBK equation with initial conditions constrained by the HERA e+p data. The x-evolution will manifest in the systematic suppression in the forward region in p+Pb reactions at the LHC energy.

To examine further the significance of the CGC approach, more exclusive observables such as e.g., the azimuthal correlations[16], and also heavy quark pair production and correlations[17, 18, 19], should be elaborated. Application to the cosmic ray events is also very intriguing. Furthermore, full NLO evaluations of the hadro-production cross-sections are necessary for theory consistency and for more quantitative investigations[20].

Acknowledgments

The author is very grateful to J. Albacete, A. Dumitru, K. Itakura and Y. Nara for fruitful collaborations on related topics. This work was partially supported by Grant-in-Aids for Scientific Research ((B) 22340064 and (C) 24540255).

References

[1] For a recent review,
F. Gelis, E. Iancu, J. Jalilian-Marian, R. Venugopalan, Ann. Rev. Nucl. Part. Sci. **60** (2010), 463.
[2] I. Balitsky, Phys. Rev. D **75** (2007), 014001.
[3] Y.V. Kovchegov, H. Weigert, Nucl. Phys. A **784** (2007), 188.
[4] J.L. Albacete, Phys. Rev. Lett. **99** (2007), 262301.
[5] K.J. Golec-Biernat,M. Wusthoff, Phys. Rev. D **59** (1998), 014017.
[6] J.L. Albacete, N. Armesto, J.G. Milhano, C.A. Salgado, Phys. Rev. D **80** (2009), 034031. J.L. Albacete, N. Armesto, J.G. Milhano, P. Quiroga Arias, C.A. Salgado, Eur. Phys. J. C **71** (2011), 1705.
[7] H.J. Drescher, Y. Nara, Phys. Rev. C **75** (2007), 034905; *ibid* **76** (2007), 041903.
[8] J.L. Albacete, A. Dumitru, arXiv:1011.5161v3[hep-ph] 2010.
[9] J. Albacete, A. Dumitru, H. Fujii and Y. Nara, Nucl. Phys. A **897** (2013) 1 [arXiv:1209.2001[hep-ph]].
[10] A. Dumitru, A. Hayashigaki, J. Jalilian-Marian Nucl. Phys. A **765** (2006), 464; *ibid* A **770** (2006), 57.
[11] J. Pumplin et al., JHEP 0207 (2002), 012.
[12] D. de Florian, R. Sassot, M. Stratmann Phys. Rev. D **75** (2007), 114010; *ibid* **76** (2007), 074033.
[13] H. Fujii, K. Itakura, Y. Kitadono and Y. Nara, J. Phys. G **38** (2011), 124125 [arXiv:1107.1333 [hep-ph]].
[14] I. Arsene *et al.* [BRAHMS Collaboration] Phys. Rev. Lett. **93** (2004), 242303.
[15] J. Adams *et al.* [STAR Collaboration] Phys. Rev. Lett. **97** (2006), 152302.
[16] J.L. Albacete, C. Marquet, Phys. Lett. B **687** (2010), 174; Phys. Rev. Lett. **105** (2010), 162301.
[17] J. P. Blaizot, F. Gelis and R. Venugopalan, Nucl. Phys. A **743** (2004), 57 [arXiv:hep-ph/0402257].
[18] H. Fujii, F. Gelis and R. Venugopalan, Phys. Rev. Lett. **95** (2005), 162002 [arXiv:hep-ph/0504047]; Nucl. Phys. A **780** (2006), 146 [arXiv:hep-ph/0603099].
[19] K. Watanabe and H. Fujii, in progress.
[20] G. A. Chirilli, B. W. Xiao and F. Yuan, arXiv:1112.1061 [hep-ph].

Unstable dynamics of Yang-Mills fields at early times of heavy ion collisions

Andreas Ipp

Institut für Theoretische Physik, Technische Universität Wien, Wiedner Hauptstraße 8-10, A-1040 Vienna, Austria

E-mail: ipp@hep.itp.tuwien.ac.at

Abstract. The quark gluon plasma as produced in heavy ion collisions is exposed to early anisotropies in momentum space due to its rapid expansion. Such anisotropies can lead to non-abelian plasma instabilities, driven by unstable gluonic modes that grow exponentially fast. These plasma instabilities can be simulated using a discretized version of gauge-covariant Boltzmann-Vlasov and Yang-Mills equations. In the non-expanding case, a turbulence cascade forms, which is associated with an approximately linear growth of energy in collective fields. Early longitudinal expansion slows down the growth of unstable modes, and the formation of soft gluonic fields depends crucially on the initial conditions assumed.

1. Introduction

Heavy ion colliders like RHIC or LHC create the quark-gluon plasma (QGP) in an anisotropic state. Due to its fast expansion, initially along the longitudinal but later also along the transverse directions, the plasma cools quickly and only exists for a duration of about a few tens of yoctoseconds (1 ys $= 10^{-24}$ s). At early times and close to the center of the QGP, longitudinal expansion dominates over transverse expansion. This quickly leads to strong momentum anisotropies along the polar angle with respect to the collision axis [1]. Such early polar momentum space anisotropies could allow for a violation of the viscosity bound [2], or lead to the emission of photon double pulses that are separated merely by yoctoseconds [3]. Most notably, polar momentum space anisotropies can induce Chromo-Weibel plasma instabilities, which are generalizations of Weibel or filamentation instabilities in ordinary electromagnetic plasmas [4]. It has been suggested early that these instabilities may play a fundamental role in the QGP [5, 6, 7]. In fact, already an infinitesimal amount of momentum space anisotropy causes the appearance of instabilities in collisionless plasmas [8, 9]. In electromagnetic plasmas, the current filamentation instability develops magnetic islands on a fast electron time scale induced by the deflection of the original electron streams, which eventually leads to an isotropization of the electron distribution [10]. This occurs on a time scale which is faster than ordinary perturbative scattering processes, and could thus explain the fast apparent isotropization and thermalization that is suggested by hydrodynamic modeling of the early QGP evolution [11, 12].

Non-abelian plasma instabilities can be studied using real-time lattice simulations [14, 15, 16]. The effective field theory for the collective phenomena at the soft scales is provided by gauge-covariant collisionless Boltzmann-Vlasov equations [17]. The corresponding effective action is nonlocal and nonlinear [18, 19], but can be made local using auxiliary fields in the adjoint

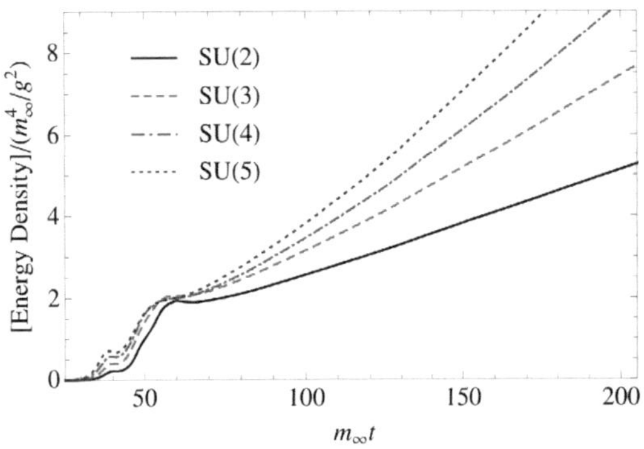

Figure 1. Comparison of average total field energy densities $\mathcal{E}$ for SU(2) through SU(5) on linear scale in 3+1 dimensional simulations for a non-expanding system with anisotropy parameter $\xi = 10$ [13].

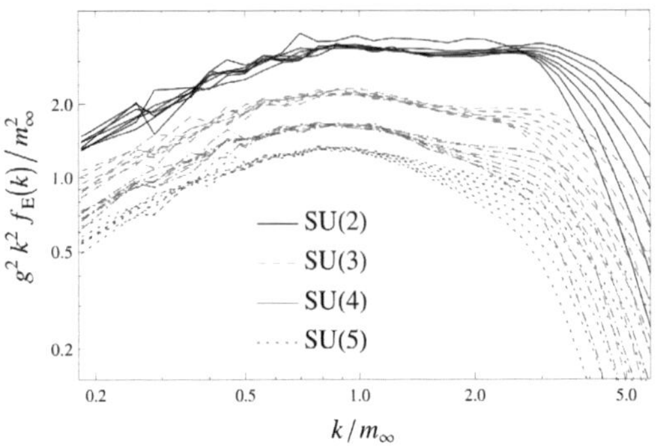

Figure 2. The power spectrum for the electric distribution $f_E(k)$ for various gauge groups SU(2) through SU(5) at late times $80 < m_\infty t < 150$. The distance between the lines is $m_\infty \Delta t \approx 11$ [13].

representation $W_\beta(x; \mathbf{v})$ which encode the fluctuations of the distribution function of colored hard particles [20]. They depend on a spatial unit vector which appears in the velocity $v^\mu = p^\mu/|\mathbf{p}|$ of a hard particle with momentum p^μ. The Yang-Mills equations are given by

$$D_\mu(A)F^{\mu\nu} = j^\nu, \tag{1}$$

where the current j^ν is calculated from the auxiliary fields

$$j^\mu[A] = -g^2 \int \frac{d^3p}{(2\pi)^3} \frac{1}{2|\mathbf{p}|} p^\mu \frac{\partial f(\mathbf{p})}{\partial p^\beta} W^\beta(x; \mathbf{v}). \tag{2}$$

The non-abelian Boltzmann-Vlasov equation for soft fields reduces to

$$[v \cdot D(A)]W_\beta(x; \mathbf{v}) = F_{\beta\gamma}(A)v^\gamma, \tag{3}$$

with $D_\mu = \partial_\mu - ig[A_\mu, \cdot]$. The scale of the hard particles drops out from these equations. An anisotropic distribution function $f(\mathbf{p})$ is obtained by deforming an isotropic distribution f_{iso} according to $f(\mathbf{p}) \propto f_{\text{iso}}(\mathbf{p}^2 + \xi p_z^2)$ with anisotropy parameter $-1 < \xi < \infty$ [8]. A discretized version of the above non-abelian gauge-covariant Boltzmann-Vlasov equations is formulated on a cubic lattice [15]. The unit sphere of velocities for the auxiliary fields is described by a discretized set of unit vectors [21, 15] or by an expansion in terms of spherical harmonics [14].

2. Numerical results

For stationary anisotropic plasmas, the exponential growth of non-abelian plasma instabilities is limited in 3+1 dimensions by non-abelian self-interactions. The exponential growth ceases when gluon self-interactions are no longer negligible at a certain magnitude of soft fields. These self-interactions lead to a turbulence cascade which form a power-law distribution $f_k \propto k^{-\nu}$ with a spectral index that turns out to be about $\nu \approx 2$. While simulations are usually based on the gauge group SU(2) [22, 16, 23], it has been confirmed that the same spectral index holds also in the QCD gauge group SU(3) as well as in higher gauge groups [13]. The systematics of the scaling with N_c in the non-abelian regime is shown in Fig. 1 where the gauge groups SU(2) through SU(5) are compared. During the exponential growth, the Chromo–Weibel instability occurs within a single color direction so that additional colors do not modify the growth rate. This changes when non-Abelian self-interactions become important. One observes that for different gauge groups the energy densities cease to grow at approximately the same value. In the following linear regime, larger gauge groups grow faster than smaller ones by roughly a factor of N_c. This indicates that the turbulence cascade picture depends crucially on the number of colors involved, since larger gauge groups allow for more color modes to be filled by the cascade. Figure 2 shows the late-time behavior of spectra for various gauge groups. The spectra are multiplied by k^2 so that a scaling with $\nu \approx 2$ would correspond to a horizontal line. The slow growth at large momenta corresponds to the linear growth regime of Fig. 1.

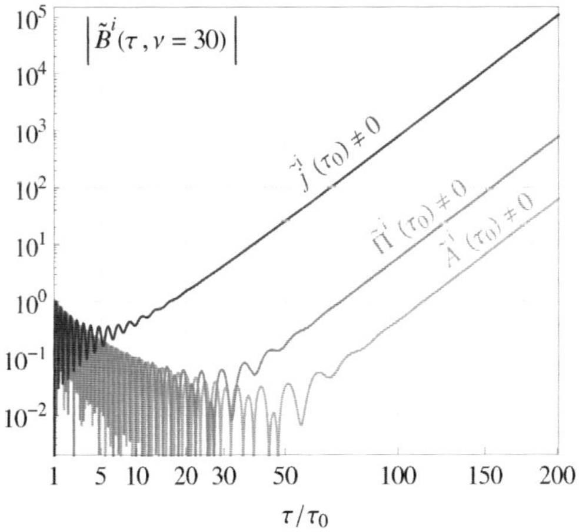

Figure 3. Magnetic Weibel instabilities on an expanding background. The influence of different initial values is studied for modes with wave vector $\nu = 30$ in the direction of the anisotropy [24].

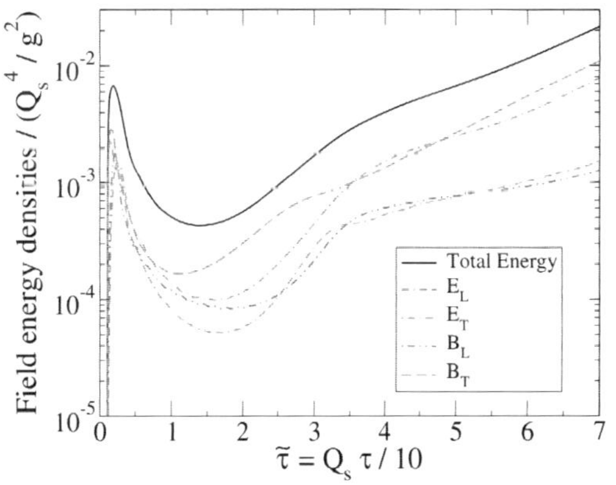

Figure 4. Chromoelectric, chromomagnetic, and total energy densities as a function of proper time. Proper time is normalized such that for $Q_s = 2$ GeV each unit of $\Delta\tilde{\tau}$ corresponds to 1 fm/c [25].

A longitudinal expansion of the plasma at early times modifies the exponential growth, as there are two competing effects: On the one hand, plasma instabilities drive the exponential growth, on the other hand the longitudinal expansion suppresses the growth. The net effect is a reduction from a growth exponential in time to exponential in the square root of proper time. This has been numerically observed using the color glass condensate scheme [26, 27] and the discretized hard loop scheme [28, 29]. In those simulations, an uncomfortable delay of the onset of plasma instabilities has been observed [29]. Collective fields decay and suppress Weibel instabilities at early times of the expanding plasma. It has been pointed out though that this suppression depends strongly on the initial conditions assumed [24]. In Fig. 3, the delay of the growth is displayed for various possibilities of initial conditions. Early work concentrated on a seed electric field with only $\tilde{\Pi}^i(\tau_0) \neq 0$ [28] or a seed magnetic field with only $\tilde{A}^i(\tau_0) \neq 0$ [29]. Seed magnetic fields instead of seed electric fields increase the delay of plasma instabilities, as do mixed initial conditions for the fields. Surprisingly, if one considers initial fluctuations in the currents with only $\tilde{j}^i(\tau_0) \neq 0$, the delay is very strongly reduced [24].

This behavior has been confirmed in numerical simulations in 3+1 dimensions. Figure 4 shows various energy densities as a function of proper time for a longitudinally expanding system. Initially, the soft fields are depleted by the longitudinal expansion. Only after some time, the unstable modes overcome the depletion and all field components reach approximately the same magnitude. The growth rate is moderately reduced and transverse chromoelectric and chromomagnetic fields begin to dominate the energy density. Contrary to the fixed-anisotropy simulations, a saturation of the roughly exponential growth is not observed [25].

Since plasma instabilities may cause the isotropization in plasmas, the question remains how to measure the isotropization time. Direct photons would be a good indicator because they leave the QGP likely without further interaction. Under certain conditions, that is non-central collisions and photon production in a direction close to forward direction, non-trivial photon pulse shapes may be expected due to intermediate non-isotropic photon emission [3]. In extreme cases, these pulse envelopes may assume the shape of double pulses at the yoctosecond time scale. While it will not be possible to resolve such time structures directly [30], it may be possible to observe the effect of such modifications to the pulse envelope through Hanbury Brown-Twiss [31] correlation measurements. A photon detector in the required forward direction may be installed during the ALICE detector upgrade by 2018 when the proposed Forward Calorimeter may be installed. With a few hundred photon pairs expected per year, such a measurement may be challenging, but not impossible. Thus, photons could provide valuable information about the earliest times of the plasma evolution, where gluon dynamics may be subject to plasma instabilities due to the rapid expansion along the beam axis.

Concluding, one can state that for heavy-ion collisions at RHIC, non-abelian plasma instabilities probably may not have enough time to develop as they compete against the fast longitudinal expansion, but depending on the initial conditions, non-abelian plasma instabilities may play an important role at LHC energies [24, 25].

References

[1] Blaizot J P, Gelis F, Liao J F, McLerran L and Venugopalan R 2012 *Nucl. Phys.* **A873** 68–80
[2] Rebhan A and Steineder D 2012 *Phys. Rev. Lett.* **108** 021601 (*Preprint* 1110.6825)
[3] Ipp A, Keitel C H and Evers J 2009 *Phys. Rev. Lett.* **103** 152301 (*Preprint* 0904.4503)
[4] Weibel E S 1959 *Phys. Rev. Lett.* **2** 83–84
[5] Mrowczynski S 1988 *Phys. Lett.* **B214** 587
[6] Mrowczynski S 1993 *Phys. Lett.* **B314** 118–121
[7] Pokrovsky Y E and Selikhov A V 1988 *JETP Lett.* **47** 12–14
[8] Romatschke P and Strickland M 2003 *Phys. Rev.* **D68** 036004 (*Preprint* hep-ph/0304092)
[9] Romatschke P and Strickland M 2004 *Phys. Rev.* **D70** 116006 (*Preprint* hep-ph/0406188)
[10] Califano F, Attico N, Pegoraro F, Bertin G and Bulanov S V 2001 *Phys. Rev. Lett.* **86** 5293
[11] Arnold P, Lenaghan J and Moore G D 2003 *JHEP* **08** 002 (*Preprint* hep-ph/0307325)
[12] Arnold P, Lenaghan J, Moore G D and Yaffe L G 2005 *Phys. Rev. Lett.* **94** 072302 (*Preprint* nucl-th/0409068)
[13] Ipp A, Rebhan A and Strickland M 2011 *Phys. Rev.* **D84** 056003 (*Preprint* 1012.0298)
[14] Arnold P, Moore G D and Yaffe L G 2005 *Phys. Rev.* **D72** 054003 (*Preprint* hep-ph/0505212)
[15] Rebhan A, Romatschke P and Strickland M 2005 *JHEP* **09** 041 (*Preprint* hep-ph/0505261)
[16] Bödeker D and Rummukainen K 2007 *JHEP* **07** 022 (*Preprint* arXiv:0705.0180 [hep-ph])
[17] Blaizot J P and Iancu E 2002 *Phys. Rept.* **359** 355–528 (*Preprint* hep-ph/0101103)
[18] Pisarski R D 1997 hep-ph/9710370
[19] Mrowczynski S, Rebhan A and Strickland M 2004 *Phys. Rev.* **D70** 025004 (*Preprint* hep-ph/0403256)
[20] Blaizot J P and Iancu E 1994 *Nucl. Phys.* **B417** 608–673 (*Preprint* hep-ph/9306294)
[21] Rebhan A, Romatschke P and Strickland M 2005 *Phys. Rev. Lett.* **94** 102303 (*Preprint* hep-ph/0412016)
[22] Arnold P B and Moore G D 2006 *Phys. Rev.* **D73** 025006 (*Preprint* hep-ph/0509206)
[23] Arnold P and Moore G D 2007 *Phys. Rev.* **D76** 045009 (*Preprint* arXiv:0706.0490 [hep-ph])
[24] Rebhan A and Steineder D 2010 *Phys. Rev.* **D81** 085044 (*Preprint* 0912.5383)
[25] Attems M, Rebhan A and Strickland M 2012 (*Preprint* 1207.5795)
[26] Romatschke P and Venugopalan R 2006 *Phys. Rev. Lett.* **96** 062302 (*Preprint* hep-ph/0510121)
[27] Romatschke P and Venugopalan R 2006 *Phys. Rev.* **D74** 045011 (*Preprint* hep-ph/0605045)
[28] Romatschke P and Rebhan A 2006 *Phys. Rev. Lett.* **97** 252301 (*Preprint* hep-ph/0605064)
[29] Rebhan A, Strickland M and Attems M 2008 *Phys. Rev.* **D78** 045023 (*Preprint* 0802.1714)
[30] Ipp A, Evers J, Keitel C H and Hatsagortsyan K Z 2011 *Phys. Lett.* **B702** 383–387 (*Preprint* 1008.0355)
[31] Ipp A and Somkuti P 2012 *Phys. Rev. Lett.* **109** 192301 (*Preprint* 1207.0197)

Journal of Physics: Conference Series **422** (2013) 012029 doi:10.1088/1742-6596/422/1/012029

Strong-field physics as a new probe of early-time dynamics in heavy-ion collisions

Kazunori Itakura

Theory Center, IPNS, KEK, 1-1 Oho, Tsukuba, Ibaraki 305-0801, Japan

E-mail: `kazunori.itakura@kek.jp`

Abstract. In high-energy heavy-ion collisions, there appear two kinds of extremely strong fields: Ordinary electromagnetic fields and non-Abelian color Yang-Mills fields called a glasma. I explain how they are created in collisions, how strong they are, and how important they are in understanding early-time "pre-equilibrium" evolution of collision events. As one of the typical strong-field "nonlinear QED" phenomena, I discuss possible relevance of the "vacuum birefringence" of a photon in probing the very early-time dynamics of heavy-ion collisions.

1. Introduction: What is strong-field physics?

One of the most difficult unsolved problems in high-energy heavy-ion collisions (HICs) is to clarify the mechanism that drives created matter into a thermalized QGP. In this talk, I emphasize that strong-field physics provides a new tool to investigate the early-time dynamics of HICs. By "strong-field physics", I mean characteristic phenomena that occur only when the (external) gauge fields are strong. For example, consider an electron propagator in a strong magnetic field (Fig. 1, left). Each insertion of an external magnetic field gives a factor eB/m_e^2 and thus such tree diagrams are enhanced when the magnetic field is stronger than the "*critical magnetic field*", $B_c \equiv m_e^2/e$, and all order summation of such diagrams yields results which are non-perturbative with respect to the coupling, and nonlinear with respect to the magnetic field B. Figure 2 shows typical strong-field phenomena that occur with such 'dressed' electrons under strong magnetic fields: (a) vacuum polarization which induces birefringence phenomena of a photon, and (b) photon splitting which is prohibited in the ordinary vacuum. Notice that the corresponding electric field $E_c = m_e^2/e$ is the "*Schwinger field*" beyond which creation of e^+e^- pairs from the vacuum becomes possible. All these kinds of phenomena are called "nonlinear QED effect" (for more details in relation to heavy-ion physics, see Ref. [1]).

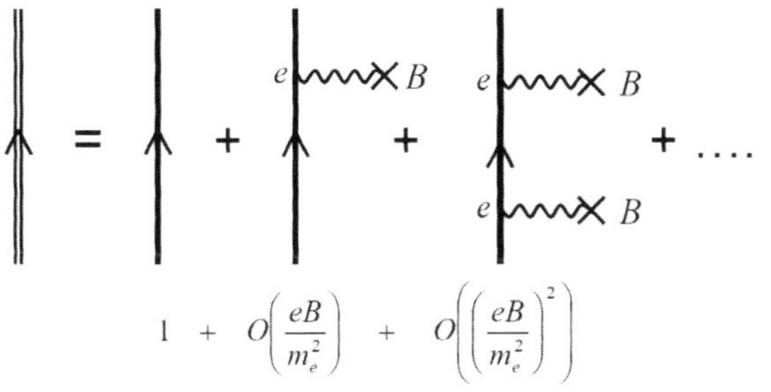

Figure 1. Electron propagator in an external magnetic field.

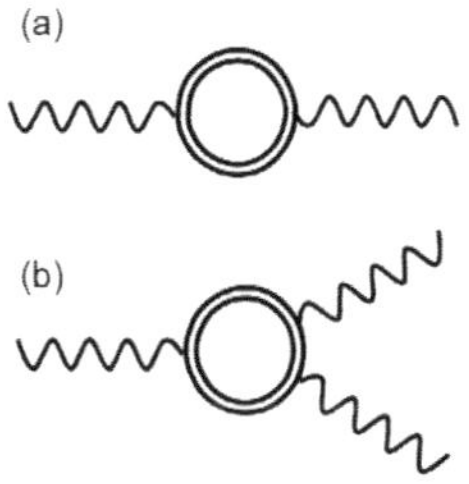

Figure 2. (a) vacuum polarization (b) photon splitting

Strength	Realized as
8.3×10^4	supercond. magnets in LHC
4.5×10^5	strongest steady field
$\sim 10^{12}$	neutron stars
$\mathbf{4\times10^{13}}$	**critical field for electrons**
$\sim 10^{15}$	magnetars
$\sim 10^{17}, 10^{18}$	noncentral HIC at RHIC & LHC

Strength	Realized as
$0.5\,(= m_e)$	critical field for electrons (EM)
$\sim 2-3$	magnetars (EM)
$\sim 10^2(\sim m_\pi)$	noncentral HIC at RHIC (EM)
$\sim 4\times 10^2$	noncentral HIC at LHC (EM)
$\sim 10^3(\sim Q_s)$	gluon fields at RHIC (YM)
$(2-3)\times10^3$	gluon fields at LHC (YM)

Table 1. Comparison of EM and YM fields in various situations in unit of Gauss (left, only ordinary magnetic field B) and MeV (right, $\sqrt{eB}$ for EM fields and $\sqrt{g\mathcal{B}}$ for YM fields).

2. Two strong fields in HICs

High-energy HICs create two different kinds of strong fields: Ordinary electromagnetic fields and non-Abelian color Yang-Mills fields called a glasma [3]. Both of them are strongest at the earliest time, but they play different roles. The glasma is a transitional state between the initial condition (color glass condensate) and QGP, and thus we need to understand thermalization as a result of time evolution of the glasma. On the other hand, the electromagnetic fields appear in noncentral collisions, and are absent (or, at most, small with fluctuation included) in central collisions where creation of QGP is easier than in noncentral collisions. Thus, the electromagnetic fields will not play a significant role in thermalization, but characteristic phenomena induced by strong electromagnetic fields can be good probes of the earliest-time matter.

Let us overview how strong they are. Table 1 is a comparison of magnetic fields realized in various situations. It shows several situations in Nature (neutron stars, magnetars), but in fact the magnetic fields created in high-energy HICs are much stronger than any of them including the critical magnetic field for electrons $B_c = m_e^2/e = 4\times10^{13}$Gauss. The maximum strength amounts to $\sim 10^{17}$Gauss at RHIC and $\sim 10^{18}$Gauss at LHC, far above the critical value. Since the origin of magnetic fields is fast moving nuclei with electric charges, such strong fields last only for a very short period (see Fig. 3). However, with the strongest "supercritical" magnetic fields, we expect many interesting phenomena related to nonlinear QED to occur.

HICs produce non-Abelian glasma fields, too. Similar to the electromagnetic fields discussed above, the origin of glasma fields is also the colliding nuclei. At high energy, the colliding nucleus is dominated by gluons having very small fractions of momentum. They are successively emitted from valence partons distributing randomly on the transverse disk (Lorentz-contracted nucleus), and are densely populated. Thus they are collectively treated as a field, and its strength is very high. This state is called the color glass condensate (CGC). After the collision, this dense gluons generate a glasma field that is also strong. Reflecting the random structure on the transverse disk in CGC, the glasma has a flux tube structure extending in the beam direction. Thus the problem is to understand time evolution of the glasma with a flux tube structure towards thermalization.

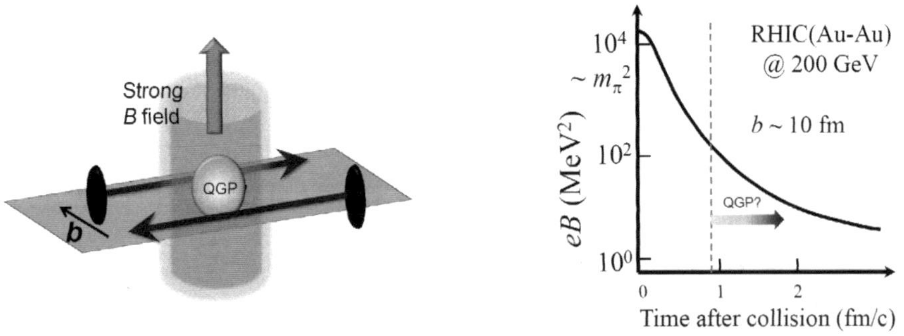

Figure 3. In noncentral HICs, very strong magnetic fields are created perpendicular to the reaction plane. They decay rapidly, but are still strong enough when QGP is formed at $\tau \sim 1$ fm [2].

Journal of Physics: Conference Series **422** (2013) 012029 doi:10.1088/1742-6596/422/1/012029

Since one cannot directly compare the strength of color fields with that of ordinary electromagnetic fields, let us compare them in unit of energy. Right of Table 1 is a comparison of the ordinary and color electromagnetic fields. For the electromagnetic fields we show $\sqrt{eB}$ in unit of MeV, and for color electromagnetic fields $\sqrt{g\mathcal{B}}$ in the same energy scale where g is the coupling strength and $\mathcal{B}$ is the color magnetic fields. Notice that the critical "color" field in QCD should be determined by quark mass m_q. Still, the strength $\sqrt{g\mathcal{B}} \sim Q_s \sim 1$ GeV at RHIC (Q_s is the saturation momentum, a typical momentum of gluons in CGC) is well beyond the critical value $m_q \sim$ a few MeV, suggesting the importance of Schwinger mechanism of $q\bar{q}$ pair production. Moreover, gluon pair production is also possible from color "electric" glasma since gluons are massless. Recently, it is gradually being recognized that unstable dynamics of the glasma is important for, at least, isotropization of the system, which should be the first step towards thermalization. In addition to the Schwinger mechanism mentioned above, the "magnetic" glasma fields are unstable against small fluctuation perpendicular to the direction of the glasma color field [4, 5]. This is a revival of the famous Nielsen-Olesen instability [6] which was originally discussed in context of finding a true vacuum in Yang-Mills theory. Most recently, it has also been found that in the presence of both the color magnetic and electric fields, anomalously large number of gluons will be produced via the Schwinger mechanism enhanced by the Nielsen-Olesen instability [7].

3. Vacuum birefringence of a photon and possible effects in HICs

Photons are emitted at all stages of HICs. Direct photons are created in the initial hard collisions, thermal photons are from QGP, and decay photons are from decays of (mainly) pions. In addition to these well-known photon sources, in the presence of strong magnetic fields, there will be synchrotron radiation from quarks [8]. Since photons do not directly interact via strong interaction, they escape from the collision area, and carry the information of initial stages of HIC events. However, when extracting such information, we have to be careful about the nonlinear QED effects on photons. In particular, the initial direct photons and thermal photons from the earliest time QGP will be affected by the magnetic field through the vacuum birefringence.

In an ordinary vacuum, the one-loop self-energy diagram of a photon propagator, or the polarization tensor $\Pi^{\mu\nu}(q)$, does not change the speed of light. However, in the presence of a strong magnetic field that selects a specific direction in space, the tensor $\Pi^{\mu\nu}(q, B)$ acquires additional terms that depend on B (see Fig. 2(a)). Consequently, a photon can have, in general, refractive indices different from unity just as if the vacuum behaves as a medium. The refractive indices of two physical propagating modes are not necessarily the same, thus this is called "birefringence". More specifically, the polarization tensor of a photon at one-loop level is given by

$$i\Pi^{\mu\nu}(q,B) = (-ie)^2(-1)\int \frac{d^4p}{(2\pi)^4}\mathrm{Tr}\left[\gamma^\mu G(p,A)\gamma^\nu G(p+q,A)\right], \qquad (1)$$

where $G(p, A)$ is the electron propagator in a magnetic field, $iG^{-1}(p, A) = \not{p} - e\not{A} - m_e$. Note that this propagator incorporates all-order contributions with respect to the external magnetic field. We assume that the magnetic field is oriented to the third direction in spatial coordinates. With the special direction specified by the magnetic field, the Lorentz structure of the polarization tensor $\Pi^{\mu\nu}$ is now modified as

$$\Pi^{\mu\nu}(q,B) = -\chi_0\left[q^2\eta^{\mu\nu} - q^\mu q^\nu\right] - \chi_1\left[q_\parallel^2\eta_\parallel^{\mu\nu} - q_\parallel^\mu q_\parallel^\nu\right] - \chi_2\left[q_\perp^2\eta_\perp^{\mu\nu} - q_\perp^\mu q_\perp^\nu\right], \qquad (2)$$

where the metric $\eta^{\mu\nu} = diag(1, -1, -1, -1)$ and the photon momentum $q^\mu = (q^0, q_\perp, 0, q^3)$ are decomposed into two parts (parallel and perpendicular to the magnetic field); $\eta_\parallel^{\mu\nu} = diag(1, 0, 0, -1)$, $\eta_\perp^{\mu\nu} = diag(0, -1, -1, 0)$, and $q_\parallel^\mu = (q^0, 0, 0, q^3)$, $q_\perp^\mu = (0, q_\perp, 0, 0)$, and χ_i

$(i = 0, 1, 2)$ are scalar functions of $q_{\parallel}^2$, $q_{\perp}^2$ and B. Without the magnetic field, we see that both χ_1 and χ_2 vanish, and that the last two terms are absent in a vacuum. With this polarization tensor, the Maxwell equation gets modified as $[q^2\eta^{\mu\nu} - q^\mu q^\nu - \Pi^{\mu\nu}(q, B)]A_\nu(q) = 0$. By solving this equation, one obtains dispersion relations for two physical modes. One finds that, when χ_1 and/or χ_2 are nonzero, these two are distinct from each other differently from those in a vacuum, $\omega^2 = \mathbf{q}^2$. Defining the refractive index n by

$$n^2 = \frac{|\mathbf{q}|^2}{\omega^2}, \tag{3}$$

one obtains two different refractive indices

$$n_1^2 = \frac{1 + \chi_0 + \chi_1}{1 + \chi_0 + \chi_1 \cos^2\theta}, \qquad n_2^2 = \frac{1 + \chi_0}{1 + \chi_0 + \chi_2 \sin^2\theta}, \tag{4}$$

where $q^\mu = (\omega, q_\perp, 0, q_\parallel) = (\omega, |\mathbf{q}|\sin\theta, 0, |\mathbf{q}|\cos\theta)$. Of course, when χ_1 and χ_2 are vanishing, both n_1 and n_2 reduce to unity. It should be noticed that the refractive indices n_i depend on the angle θ, and when $\theta = 0$ (a photon propagates in parallel to the magnetic field), both n_1 and n_2 are 1. Most recently, one-loop contribution to the vacuum polarization tensor has been analytically computed [9], and one can explicitly compute the refractive indices in strong magnetic fields.

As mentioned above, there is an azimuthal angle θ dependence. When a photon propagates in parallel to the magnetic field, or perpendicularly to the reaction plane in noncentral HIC, there is no effect. On the other hand, if a photon propagates perpendicularly to the magnetic field, or in the reaction plane, then modification of refractive indices is very large. If the photon energy is large enough, it can even decay into an e^-e^+ pair. Therefore, if the photons are emitted isotropically, magnetic fields will add azimuthal dependence (leading to negative v_2).[1] Besides, since the magnetic field should have nontrivial profile, if trajectories of two photons are different, the effects of birefrincence will be different. Thus, we expect distortion of the HBT image in strong magnetic fields. This was studied in a very simple model [10], and it was found that the strong magnetic field works like a distorted lens. From these nonlinear QED effects, we are, in principle, able to extract information of the early-time HIC events. I hope future experiments can resolve these physics and contribute to disentangle the early-time dynamics of HIC events.

References

[1] K. Itakura, in "Proceedings of International Conference on Physics in Intense Fields (PIF2010)," edited by K. Itakura, et al., available from `http://ccdb5fs.kek.jp/tiff/2010/1025/1025013.pdf`

[2] D. E. Kharzeev, L. D. McLerran and H. J. Warringa, Nucl. Phys. A **803** (2008) 227, V. Skokov, A. Y. Illarionov and V. Toneev, Int. J. Mod. Phys. A **24** (2009) 5925-5932, A. Bzdak and V. Skokov, Phys. Lett. B **710** (2012) 171, W. T. Deng and X. G. Huang, Phys. Rev. C **85** (2012) 044907.

[3] T. Lappi and L. McLerran, Nucl. Phys. A772 (2006) 200.

[4] H. Fujii and K. Itakura, Nucl. Phys. A809 (2008) 88, A. Iwazaki, Prog. Theor. Phys. **121** (2009) 809,

[5] P. Romatschke and R. Venugopalan, Phys. Rev. Lett. **96** (2006) 062302; Phys. Rev. D **74** (2006) 045011, K. Fukushima and F. Gelis, Nucl. Phys. A **874** (2012) 108, J. Berges, S. Scheffler, S. Schlichting and D. Sexty, Phys. Rev. D **85** (2012) 034507

[6] N. K. Nielsen and P. Olesen, Nucl. Phys. **B144** (1978) 376.

[7] N. Tanji and K. Itakura, Phys. Lett. B **713** (2012) 117.

[8] K. Tuchin, Phys. Rev. C **82** (2010) 034904.

[9] K. Hattori and K. Itakura, "*Vacuum birefringence in strong magnetic fields: (I) Photon polarization tensor with all the Landau levels*", `arXiv:1209.2663 [hep-ph]`, "*Vacuum birefringence in strong magnetic fields: (II) Complex refractive index in the lowest Landau level approximation*", in preparation.

[10] K. Itakura and K. Hattori, PoS WPCF **2011** (2011) 042.

[1] However, we need to also consider synchrotron radiation from quarks which will contribute positively to v_2. Synchrotron radiation is emitted along the circular trajectory of the quark.

Journal of Physics: Conference Series **422** (2013) 012030 doi:10.1088/1742-6596/422/1/012030

Physics at low x_{Bj} and in eA at the LHeC

Néstor Armesto for the LHeC Study Group

Departamento de Física de Partículas and IGFAE, Universidade de Santiago de Compostela, 15782 Santiago de Compostela, Galicia, Spain

E-mail: nestor.armesto@usc.es

Abstract. I discuss the project of an electron-proton/ion collider at CERN using the LHC beams, the Large Hadron Electron Collider. After a brief introduction, I present the project. Then I summarize the possibilities for small-x studies and for electron-nucleus collisions. I also indicate the status of the project and future steps to be taken.

1. Introduction

From studies at fixed-target experiments and at the HERA accelerator at DESY, it has been concluded that inclusive and diffractive data at small-x can be described by non-perturbative models and, more interestingly, by different realizations of evolution equations within perturbative QCD: the standard explanation within fixed-order perturbation theory (DGLAP evolution equations), resummation schemes, and non-linear approaches. Concerning the last item, unitarity of QCD as a quantum field theory implies that non-linear phenomena are unavoidable and saturation of partonic densities is expected to occur at high energies or small Bjorken-x. The Color Glass Condensate offers a non-perturbative but weak coupling realization of saturation ideas, see [1] and references therein. The present discussion lies on the relevant kinematical regime for such phenomena and on the possibilities offered by existing or future experimental data to distinguish among the different available schemes.

Even ignoring such fundamental considerations and focusing merely on practical purposes, our knowledge of the gluon distribution at small x both in protons and nuclei does not suffice for the required precision in predictions within collinear factorization at hadron colliders. Besides, in the lepton-nucleus case and in the semihard region for particle production, collinear factorization is not expected to work and other factorization schemes have been proposed. Both aspects are of great importance for the study of hadronic and nuclear collisions.

The Large Hadron-electron Collider (LHeC [2]) is an electron-proton/ion collider currently under design at CERN, which will collide $20 \div 140$ GeV $e^{\pm}$ against the LHC beams, with a goal luminosity 10^{33} cm^{-2}s^{-1}. The machine is subject, apart from several requirements on detector performance, to the constraints of creating minimal disturbance during construction and working simultaneously to the LHC. Besides electro-weak studies and searches for new physics, this machine will perform precision QCD studies and it should allow an unambiguous access to the novel regime of QCD in which unitarity constraints are at work - the dense region shown in Fig. 1. With the transition between the dilute region and the new phase being a density effect, a two-pronged approach will be pursued: either decreasing x at fixed mass number A and Q^2, or increasing A at fixed x and Q^2. The LHeC will give access to a completely new region of the Q^2-x plane, see Fig. 2 for the eA case.

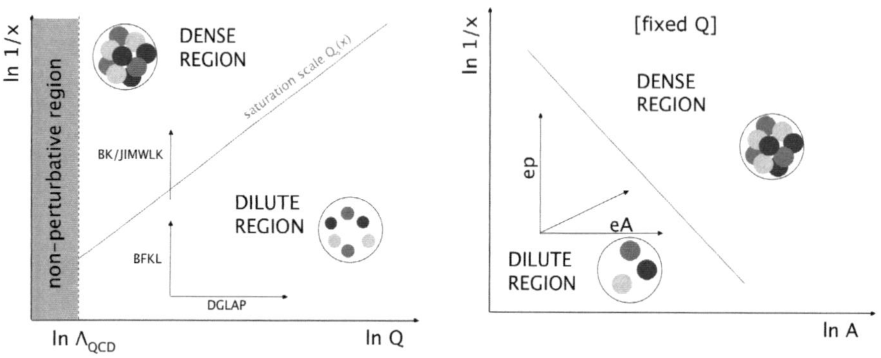

Figure 1. Sketch of the access to the dense partonic region where unitarity effects are essential, from the dilute one where linear evolution is valid. See the text for explanations. (From [2].)

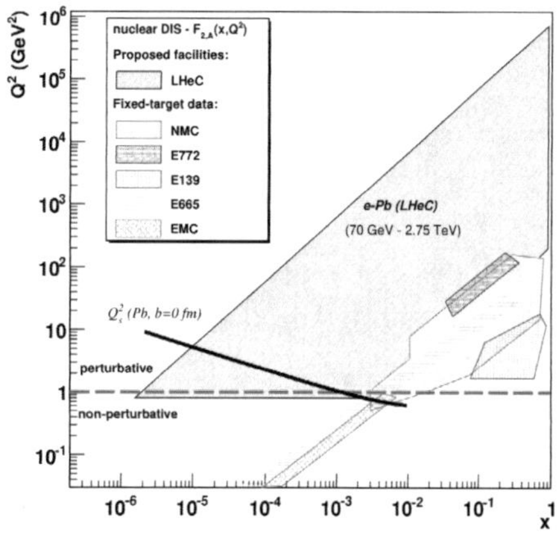

Figure 2. Region of the Q^2-x plane that will be explored with the LHeC in ePb, compared to those achievable at existing eA experiments. An estimation of the saturation scale indicating the dilute-dense transition is shown. (From [2].)

In this contribution I will mention some aspects of the QCD studies that may be performed at the LHeC. Due to the limitations of space, I will focus on small-x physics and nuclear targets, which can be addressed with the LHeC. More information can be found in [2], and in related work concerning the proposed Electron-Ion Collider in the USA [3]. After the completion of the Conceptual Design Report [2], the next steps are the submission for consideration in the European Strategy for Particle Physics [4], and a Technical Design Report for 2015.

2. Inclusive observables at small x

The LHeC will give access to a completely new region of the Q^2-x plane, see Fig. 2. With this huge kinematical lever arm and the possibility to measure not only the total structure function, F_2, but also its flavor decomposition and the longitudinal one, F_L, (see Fig. 3 for examples of LHeC pseudodata on nuclear ratios of F_2 and F_L), the LHeC offers huge possibilities for:

• Constraining the parton density functions (PDFs) in DGLAP analysis, both in ep and eA (see [2]), particularly for sea quarks and gluons. For this purpose, the combination of F_2, F_L and

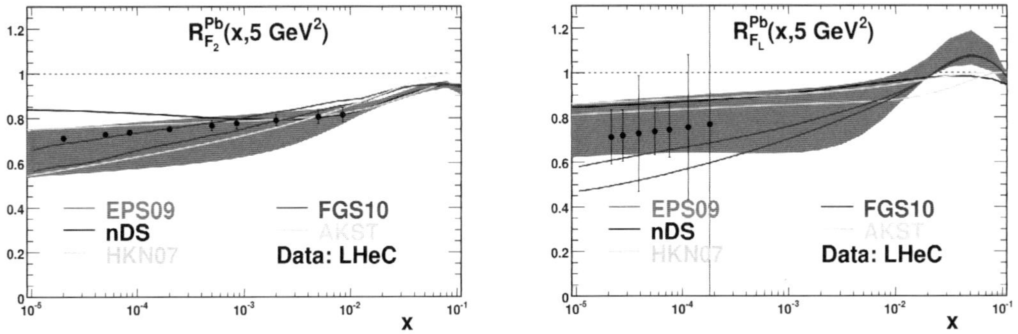

Figure 3. Predictions from different models for the nuclear ratio $R^{\rm Pb}_{F_i} = F_i^{\rm Pb}/(208\, F_i^{\rm p})$, $i = 2, L$, at small x, see the legend on the plots. Filled circles are LHeC pseudodata. (From [2].)

$F_{2c,b}$ appears to be very promising. As shown in Fig. 3 for F_2 and F_L in the nuclear case, the expected uncertainty of F_2 data is much smaller than the spread of existing models.
• Disentangling fixed-order evolution schemes from resummation or non-linear ones, see [5]. In this respect, the combination of data on F_2 and F_L is required.

3. Diffractive observables

On diffraction[1] [2], the LHeC will explore a new domain of very low β (e.g. down to a few 10^{-4} for $Q^2 \sim 4$ GeV^2 at $x_P = 0.003$, two orders of magnitude smaller than at HERA, for p; for nuclei, information on diffraction at small x does not exist). Several aspects can be highlighted:
• It will give access to diffractive masses as large as 200 GeV, providing data to check models describing the transition from low to high M_X, and to constrain nucleon and nuclear diffractive PDFs in DGLAP analyses.
• For elastic vector meson production or deeply virtual Compton scattering, a huge lever arm in W will be explored (e.g. up to ≈ 1.2 TeV for ep with $E_e = 50$ GeV) with enough precision to disentangle linear evolution schemes from non-linear ones. Besides, the differential spectrum in t will be accessible down to $t \sim -2$ GeV^2. All this will also constrain quark and gluon generalized parton densities (GPDs) at small x where they are presently unknown. Measurements of diffractive production of two vector mesons will allow the determination of helicity-flip GPDs.
• Gribov relation between diffraction in ep and nuclear shadowing will be checked in a single experimental setup (see e.g. the FGS10 and AKST models in Fig. 3, the others being NLO DGLAP analyses).
• Diffractive dijets will be measured in DIS with E_T as large as 25-30 GeV, providing a check of collinear factorization for diffractive processes.

The experimental challenge in separating inclusive diffraction ($e + {\rm A} \to e + X + X'$ with a rapidity gap) from coherent ($e+{\rm A} \to e+X+{\rm A}$) and incoherent ($e+{\rm A} \to e+X+Z{\rm p}+[A-Z]{\rm n}$) is under study.

4. Final states

The LHeC will offer huge possibilities for clarifying the dynamics of QCD radiation and hadronization. For example:

[1] The diffractive kinematical variables are $x_P = (M_X^2 - t + Q^2)/(W^2 + Q^2)$ and $\beta = x/x_P$, with M_X the diffractive mass, t the transverse momentum squared and W the γ^*-hadron center-of-mass energy.

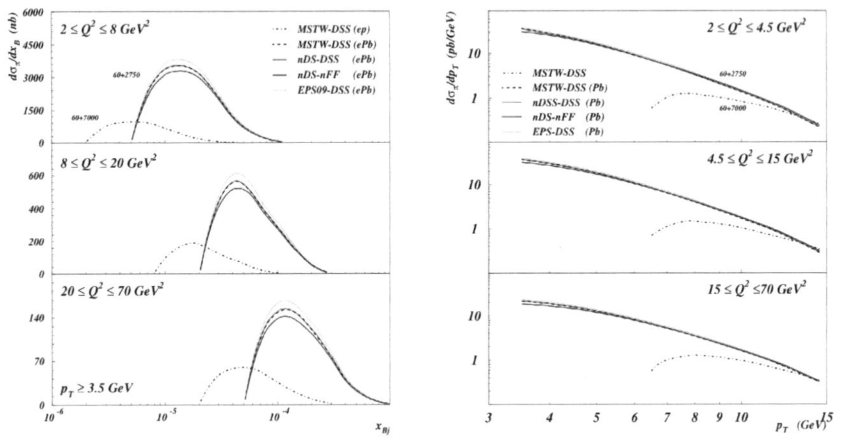

Figure 4. Cross section for inclusive π^0 production versus Bjorken x_{Bj} for $p_T > 3.5$ GeV/c (left) and versus p_T (right), computed in NLO QCD [6]. Dashed-dotted black lines refer to ep collisions. All other line types refer to ePb collisions: dashed black ones to standard nucleon PDFs and fragmentation functions, solid red and green ones to nuclear PDFs and nucleon fragmentations functions, and solid blue ones to nuclear PDFs and nuclear fragmentation functions. All cross sections are given per nucleon i.e. divided by 208 for Pb. Cuts: $\theta_\pi \in [5°, 25°]$, $x_\pi = E_\pi/E_p > 0.01$, have been applied. (From [2].)

• The dynamics of QCD radiation at small x will be studied through forward jet and particle production, which will be abundant, see Fig. 4.
• The parton/hadron energy loss mechanism in semi-inclusive DIS will be tested by introducing a piece of colored material - the nucleus - which would modify the hadronization pattern i.e. its dependence on the traversed length (by varying either the impact parameter or the nuclear size), its chemical composition,... Energies as high as 10^5 GeV in the rest frame of the nucleus will be accessible and the transition from moderate to high energies will be studied.

As an example of the abundant yield of high-energy probes, inclusive jet rates for $Q^2 = 0$ around 10^3 jets per GeV per year are expected [2] with $E_{Tjet} \sim 95$ (80) GeV in ep (ePb).

Acknowledgments

I thank the organizers for the invitation to such a nice meeting. I gratefully acknowledge financial support by the ERC grant HotLHC ERC-2001-StG-279579; by Ministerio de Ciencia e Innovación of Spain grants FPA2008-01177, FPA2009-06867-E and Consolider-Ingenio 2010 CPAN CSD2007-00042; by Xunta de Galicia grant PGIDIT10PXIB 206017PR; and by FEDER.

References

[1] F. Gelis, E. Iancu, J. Jalilian-Marian and R. Venugopalan, Ann. Rev. Nucl. Part. Sci. **60** (2010) 463 [arXiv:1002.0333 [hep-ph]].
[2] J. L. Abelleira Fernandez *et al.* [LHeC Study Group Collaboration], J. Phys. G **39** (2012) 075001 [arXiv:1206.2913 [physics.acc-ph]]; `http://cern.ch/lhec`.
[3] The Electron Ion Collider Working Group Collaboration, C. Aidala *et al.*, *A High Luminosity, High Energy Electron Ion Collider*; `http://web.mit.edu/eicc/`; D. Boer *et al.*, arXiv:1108.1713 [nucl-th].
[4] LHeC Study Group, "A Large Hadron Electron Collider at CERN", contribution no. 147 to the Open Symposium of the European Strategy Preparatory Group (10-12 September 2012, Krakow, Poland).
[5] J. Rojo and F. Caola, arXiv:0906.2079 [hep-ph].
[6] A. Daleo, D. de Florian and R. Sassot, Phys. Rev. D **71** (2005) 034013 [hep-ph/0411212].

J/ψ polarization in pp collisions with ALICE

Livio Bianchi for the ALICE Collaboration
Istituto Nazionale di Fisica Nucleare (INFN) - Torino
E-mail: lbianchi@to.infn.it

Abstract. The ALICE experiment at the LHC has studied J/ψ polarization in pp collisions at $\sqrt{s} = 7$ TeV, measuring the anisotropies in the angular distribution of the muons coming from J/ψ decay. The study was performed in the forward rapidity region ($2.5 < y < 4$), differentially in transverse momentum. In this work, the analysis is reviewed and the results are presented and compared to new theoretical predictions. Perspectives for the polarization measurement in Pb-Pb collisions are also discussed.

1. J/ψ hadroproduction and the importance of polarization

After almost fourty years from the first observation of charmonium, the J/ψ hadroproduction mechanism is still a puzzling issue. Given the high mass of the c quark, charmonium production involves different energy scales: the formation of the quark-antiquark pair is a hard process and can be reliably described through perturbative QCD calculations, but the dynamics of the bound state formation and evolution is intrinsically non-relativistic and involves soft energy scales. For this reason, no full-QCD description of quarkonium production can be carried out and theoretical models have been developed. Among them, the effective field theory NonRelativistic QCD (NRQCD), at LO [1], was found to reproduce the cross-sections measured at the Tevatron, which were significantly underestimated by Color Singlet Model (CSM) calculations at LO [2]. Nevertheless, the prediction of the J/ψ degree of polarization was contradicted by a measurement by CDF [3] and new calculations at NLO of both NRQCD and CSM [4, 5] tend to slightly mitigate but not to solve the discrepancy, that still remains an issue on which the Large Hadron Collider (LHC) is expected to provide a valuable contribution.

2. The ALICE experiment and the forward muon spectrometer

The ALICE experiment [6] is based on a central barrel, covering the pseudorapidity region $|\eta| < 2$, and a muon spectrometer, with rapidity coverage $2.5 < y_\mu < 4$. The polarization results presented here refer to inclusive J/ψ production, measured via the J/$\psi \rightarrow \mu^+\mu^-$ decay channel in the muon spectrometer. The spectrometer consists of a 10 interaction length (λ_I) thick front absorber, to remove hadrons, followed by a 3 T·m dipole magnet. Charged particles which exit the front absorber are tracked in a detector system made up of five stations, each one with two planes of cathode pad chambers. The tracking system is followed by a 7.2 λ_I iron wall, which absorbs secondary hadrons escaping the front absorber and low-momentum muons. Finally, a trigger system, based on resistive plate chambers, is used to select candidate muons with a transverse momentum larger than a given programmable threshold.
Vertex reconstruction is provided by the Inner Tracking System (ITS), an array of six layers of silicon detectors covering mid-rapidity.

Journal of Physics: Conference Series **422** (2013) 012031 doi:10.1088/1742-6596/422/1/012031

3. J/ψ polarization in pp collisions: analysis technique and results

J/ψ polarization was studied by the ALICE Collaboration [7], at forward rapidity, through the analysis of the angular distribution of the decay products, which follows the general formula:

$$W(\cos\theta,\phi) = \frac{d^2N}{d\cos\theta\, d\phi} \propto \frac{1}{3+\lambda_\theta} \cdot \left(1 + \lambda_\theta \cos^2\theta + \lambda_\phi \sin^2\theta \cos 2\phi + \lambda_{\theta\phi} \sin 2\theta \cos\phi\right), \qquad (1)$$

where θ and ϕ are the polar and azimuthal angles of one of the two muons with respect to a given reference frame [8]. The parameters λ_θ, λ_ϕ and $\lambda_{\theta\phi}$ quantify the degree of polarization; in particular $\lambda_\theta > 0$ indicates transverse polarization and $\lambda_\theta < 0$ means longitudinal polarization. Two different definitions of the polarization axis were considered:

- helicity (HE), defined as the direction of the J/ψ in the pp collision rest frame;
- Collins-Soper (CS), the bisector of the angle between the directions of one incoming proton and the opposite of the other incoming proton.

Data collected during 2010, approximately corresponding to 100 nb^{-1} of integrated luminosity, were used after the application of some standard selection criteria: events with at least one vertex reconstructed in the ITS where retained if they contained at least two tracks in the muon spectrometer, out of which at least one had to satisfy the trigger condition (1 GeV/c p_T threshold). Moreover, tracks must satisfy the condition $2.5 < y < 4$ and must also have $17.6 < R_{abs} < 88.9$ cm, where R_{abs} is the radial distance of the track from the beam axis at the downstream end of the front absorber. The latter requirement eliminates tracks passing through high-Z material regions of the absorber which are strongly affected by multiple scattering.
The two-dimensional acceptance and efficiency maps presented in Figure 1, obtained with a Monte Carlo simulation based on a parametric generator [9], show the decrease of the detector acceptance for low p_T and large $\cos\theta$ in the helicity reference frame. A similar behaviour is found for the Collins-Soper reference frame. In order to avoid very low acceptance regions, an additional cut on the p_T of the dimuons, $p_T > 2$ GeV/c, was applied and the analysis was performed in the $|\cos\theta| < 0.8$ region. For statistics reasons the study was also restricted to $p_T < 8$ GeV/c. After implementing all these cuts (6800 ± 140) J/ψ candidates survive for the analysis.

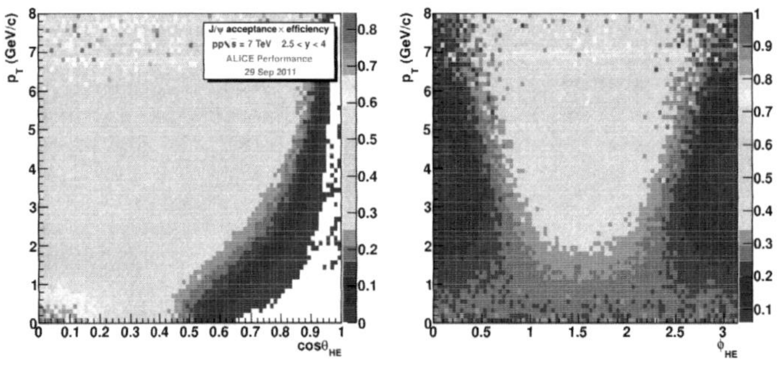

Figure 1. Acceptance times efficiency ($A \times \epsilon$) maps from Monte Carlo simulation [9] in the helicity reference frame for pp collisions at $\sqrt{s} = 7$ TeV: $\cos\theta$-p_T (left), ϕ-p_T (right).

In principle, the values of the parameters could be extracted by means of a fit to the acceptance-corrected ($\cos\theta$-ϕ) 2D distributions. However, the limited J/ψ statistics made a 2D binning impossible and the study of the angular distributions was separately performed on the polar and azimuthal variables. In particular, λ_θ and λ_ϕ were extracted from the study of

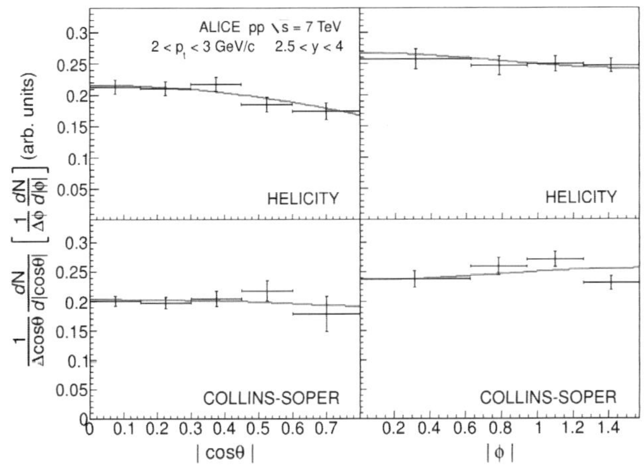

Figure 2. [7] $A \times \epsilon$ corrected angular spectra for the $2 < p_{\mathrm{T}}^{J/\psi} < 3$ GeV/c bin and $2.5 < y^{J/\psi} < 4$. The red line represents the simultaneous fit.

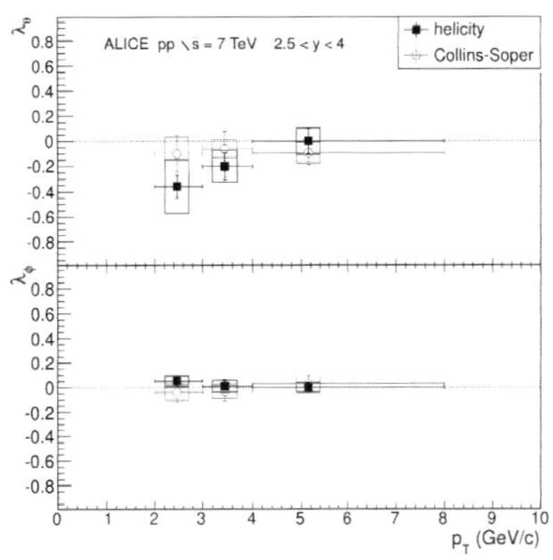

Figure 3. [7] Polarization parameters λ_θ and λ_ϕ in the helicity and Collins-Soper frames for inclusively produced J/ψ as a function of their transverse momentum (p_{T}).

the distributions

$$\mathrm{W}(\cos\theta) \propto \frac{1}{3+\lambda_\theta} \times \left(1 + \lambda_\theta \cos^2\theta\right); \qquad \mathrm{W}(\phi) \propto 1 + \frac{2\lambda_\phi}{3+\lambda_\theta}\cos 2\phi, \tag{2}$$

obtained by integrating Eq. 1 in the ϕ and $\cos\theta$ variables, respectively. In this approach the $\lambda_{\theta\phi}$ parameter was assumed to be zero and this assumption was found to be realistic with an a-posteriori check [9].

The polarization parameters were obtained by correcting the number of signal events for each bin in the angular variables for the product $A \times \epsilon$ of acceptance times detection efficiency, calculated via MC simulation [9], and then fitting the angular distributions with the functions shown in Eq. 2. The distributions of the angular variables to be used to perform the MC simulation are not known a priori, but rather represent the outcome of the data analysis: for this reason an iterative procedure was followed, reaching convergence in at most 3 steps [9].

A simultaneous study of the J/ψ polarization variables in several reference frames is particularly interesting since consistency checks on the results can be performed, using combinations of the polarization parameters which are frame-invariant [8]. In particular we made use of the invariant $F = (\lambda_\theta + 3\lambda_\phi)/(1 - \lambda_\phi)$, performing a simultaneous fit of the $|\cos\theta|$ and $|\phi|$ distributions in the two reference systems and further constraining the fit by imposing F to be the same in the CS and HE frames. In Fig. 2 we present, as an example, the result of such a fit for $2 < p_{\mathrm{T}} < 3$ GeV/c.

The final results are shown in Figure 3: in both frames, all the parameters are compatible with zero, with a possible hint for a longitudinal polarization at low p_{T} (at a 1.6σ level) in the HE frame. The comparison with recent theoretical calculations at NLO in the NRQCD and CSM frameworks, reported in [10, 11], show a better (although not perfect) agreement with NRQCD, in particular in the Collins-Soper reference frame. A conclusive comparison of data with theory requires the experiments to probe higher J/ψ transverse momenta and, in ALICE, this will be possible exploiting the higher statistics data sample collected during 2011 and 2012 (more than ten times in terms of J/ψ with respect to 2010).

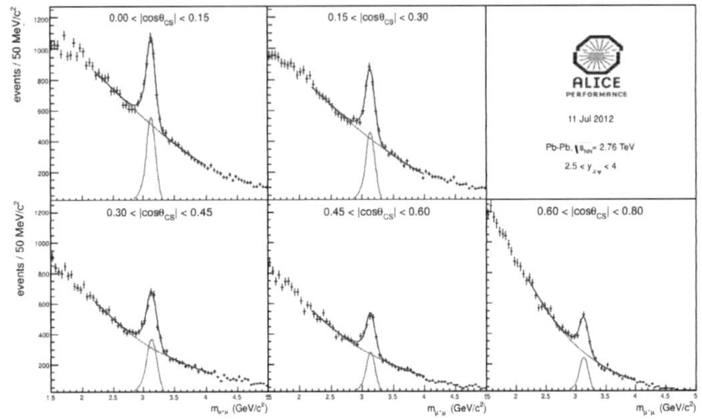

Figure 4. Invariant mass spectra for dimuons produced at forward rapidity in Pb-Pb collisions at $\sqrt{s_{NN}} = 2.76$ TeV. Five bins of $\cos\theta$ in the CS frame in the region $2 < p_{\rm T} < 8$ GeV/c.

4. Perspectives for the study of polarization in Pb-Pb collisions

The study of the degree of polarization of J/ψ produced in Pb-Pb collisions is particularly interesting for two reasons. On the one hand, all the results published by ALICE (and the other LHC experiments) on J/ψ production in heavy-ion collisions were obtained assuming no polarization. This assumption has to be carefully checked since, if not confirmed, the acceptance estimate must include polarization effects. On the other hand, a value of the parameters sensibly different from zero would reveal interesting mechanisms at play in nuclear collisions and not in pp, where there is basically no polarization.

Data collected during 2011 by ALICE in Pb-Pb collisions at $\sqrt{s_{NN}} = 2.76$ TeV are shown in Figure 4: statistics are sufficient to divide the sample in bins of angular variables and, therefore, to extract the polarization parameters. The analysis of these data is currently in progress.

5. Conclusions

ALICE has measured the polarization parameters λ_θ and λ_ϕ for inclusive J/ψ production in $\sqrt{s} = 7$ TeV *pp* collisions at the LHC. The measurement was carried out in the kinematical region $2.5 < y^{J/\psi} < 4$, $2 < p_{\rm T}^{J/\psi} < 8$ GeV/c. The polarization parameters are consistent with zero, in both the helicity and Collins-Soper reference frames. The comparison with NLO calculations of CSM and NRQCD shows a better qualitative agreement with the latter, but a firm conclusion will be only possible when higher $p_{\rm T}$ is reached by the experiments.
With the statistics collected during the Pb-Pb run of 2011, ALICE will likely be able to perform the same analysis for heavy-ion collisions.

References

[1] Cho P and Leibovich A K 1996 *Phys. Rev.* D **53** 150-162
[2] Martin A D, Glover E W N and Stirling W J 1988 *Z. Phys.* C **38** 473
[3] CDF Collaboration 1997 *Phys. Rev. Lett.* **79** 578
[4] Butenschoen M and Kniehl B A 2011 *Phys. Rev. Lett.* **106** 022003
[5] Lansberg J P 2011 *arXiv:*1107.0292
[6] ALICE Collaboration 2008 *JINST* **3** S08002
[7] ALICE Collaboration 2012 *Phys. Rev. Lett.* **108** 082001
[8] Faccioli P, Lourenço C, Seixas J and Wöhri H 2011 *Phys. Rev.* D **83** 056008
[9] Bianchi L 2012 *CERN-THESIS-2012-035*
[10] Butenschoen M and Kniehl B A 2011 *Phys. Rev. Lett.* **108** 172002
[11] Gong B, Wan L-P, Wang J-X and Zhang H-F 2012 *arXiv:*1205.6682

Summary Talk

Néstor Armesto

Departamento de Física de Partículas and IGFAE, Universidade de Santiago de Compostela, 15782 Santiago de Compostela, Galicia, Spain

E-mail: nestor.armesto@usc.es

Abstract. A brief summary of the contributions to the conference *Heavy Ion Collisions in the LHC Era*, held in Quy Nhon, Vietnam, from July 16th to 20th 2012, is provided together with some personal comments. Contributions are classified according to their link to the different stages of a high-energy heavy-ion collision.

1. Introduction

This contribution aims at providing a concise summary, together with some personal comments, of the talks presented in the conference *Heavy Ion Collisions in the LHC Era*, held in Quy Nhon, Vietnam, from July 16th to 20th 2012. The classification of the 38 talks according to different subjects, related with the different stages of a heavy-ion collision, is shown in Fig. 1. Of them, 14 are experimental talks and 24 theoretical ones. All in all, they provide an nice overview of the whole field of Heavy-Ion Physics at high energies. I will refer to the different contributions by the family name of their presenters. The overview presentation by *Blaizot* will not be summarized.

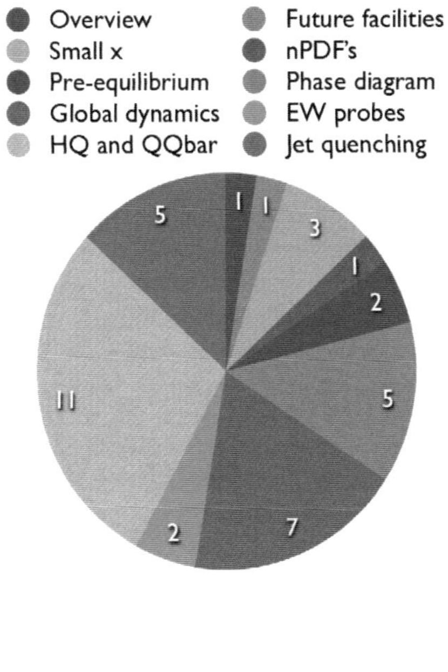

Figure 1. Distribution of the talks in subjects.

2. Initial state and pre-equilibrium

The understanding of the initial conditions for a heavy-ion collision and the path towards a collective behaviour is of major importance in the field. Equally important is the matter of the existence of a new regime of QCD in which non-linear dynamics is dominant - usually named saturation regime, for which the CGC framework has been developed. All these aspects have been treated in the conference.

2.1. Small-x physics

Nayak presented a brief overview of the status of small-x physics in hadronic and nuclear collisions, focusing mainly on the successful description of multiplicities, and nuclear suppression factors and two-particle correlations at forward rapidities. He showed the upgrade plans of ALICE, focused on the forward calorimeter FoCal that would be installed during LS2 (around 2017) and provide forward coverage (up to $\eta = 5.5$) for jets, π^0's and γ's.

Fujii presented the most recent developments in the CGC to compute single inclusive particle production, and specifically discussed the predictions for pPb collisions at the LHC with the rcBK Monte Carlo. Predictions from this approach are found to differ from those of collinear factorization only in the forward region. Uncertainties in the formalism, both from the order in the perturbative calculations and from the treatment of impact parameter, were addressed.

Beuf revised the theoretical status of the calculation of evolution equations in the CGC. By close examination of the kinematics in the higher order corrections, he proposed a simple modification to account for finite energy corrections. This is an important step towards a more reliable phenomenology in the field.

2.2. Nuclear parton densities

Helenius showed a way to introduce impact parameter dependence in any set of nuclear parton densities by expanding the nuclear dependence in powers of the profile function of the nucleus. This is a long waited step, badly needed for phenomenology. He showed the results of such procedure for EPS09. Noticeably, the results for nuclear modification factors for minimum bias and for central collisions are found to be quite close.

Armesto presented the proposal of an electron-ion collider at CERN in the TeV regime, the Large Hadron Electron Collider that implies building an electron accelerator. He showed the the large capabilities of such a machine for determining nuclear parton densities, establishing the QCD dynamics at small x and other aspects of interest for high-energy heavy-ion collisions.

2.3. Pre-equilibrium dynamics

Itakura discussed the importance of high-intensity fields in the different areas of physics. He proposed that high magnetic fields created in non-central heavy-ion collisions would induce vacuum birefringence, which could generate elliptic flow and higher harmonics in the azimuthal distributions, and distorted images in interferometry, of photons.

Ipp discussed the role of the high QCD fields created in the collision in the fast isotropisation required for hydrodynamical models to describe the experimentally observed features of data like transverse momentum and azimuthal distributions. He showed how anisotropies in the initial state lead to instabilities and the evidences that these instabilities, in the case of an expanding system, quickly isotropise it.

3. Phase diagram

During recent years, impressive progress in lattice calculations and their comparison with experiment has been done, together with new ways of exploring the phase diagram, both theoretically and experimentally, in search of its phase structure. All these aspects have been covered in the conference.

Journal of Physics: Conference Series **422** (2013) 012032 doi:10.1088/1742-6596/422/1/012032

Gupta reported on the recent, impressive progresses of lattice computations in characterizing the phase diagram of QCD. The search of the critical point through the study of moments of distributions of different particles has made possible a first direct contact between lattice and experiment. The importance of finite size corrections was also discussed.

On the other hand, *Zhuang* showed how effective models can be used to explore the region of large baryon density and the connections that appear with condensed matter physics. Specifically, he presented a NJL model with isospin chemical potential, and the resulting rich structure with a BEC→BCS transition.

Asakawa discussed the formalism of moments and its connection with lattice quantities, and the uncertainties in such connection. Concretely, he addressed the fact that in lattice baryon number fluctuations are studied, while experimentally protons are measured. He concluded that final state interactions via resonances do not spoil the connection.

Then two experimental talks were devoted to the search of the critical point. *Blume* reviewed the searches of the critical point via energy and size scan at the SPS (in NA45): particle ratios and their fluctuations versus energy, and multiplicity fluctuations. With a strong ongoing debate on how conclusive the observations are for indicating or ruling out the existence of the critical point, he also presented the future prospects of these searches in NA61 at the SPS or CBM at FAIR.

Finally, *Llope* reviewed the results of the systematic beam-energy scan (BES) at RHIC from 7.7 to 200 GeV per nucleon and the lack of evidences of a non-monotonic behaviour of the cumulants of particle distributions. The future prospects to set these conclusions on firmer ground, lie in increasing statistics and making a finer scan in energies.

4. Global dynamics

An impressive amount of data on global dynamics, particularly on particle azimuthal distributions both from RHIC and the LHC, is becoming available. All these data, reported at the conference, challenge current global models for heavy-ion collisions. New developments, both at qualitative and quantitative level, were also presented.

Yoo revised the ALICE results on multiplicities, which show agreement with expectations from saturation models, and on azimuthal distributions. Data on many Fourier moments and for identified particles thanks to ALICE unique PID capabilities, are now available up to quite large transverse momentum. They strongly indicate the need of refined event-by-event initial conditions for collective evolution.

Shi revised the results on elliptic flow of the BES at STAR, in search of non-monotonic behaviours to define the phase boundary or critical points. The main conclusion are that the difference between particles and antiparticles (showing breakdown of scaling in the number of constituent quarks) decreasing with increasing beam energy is suggestive of the importance of hadronic interactions at the lowest studied energies. ϕ elliptic flow shows a non-monotonic behaviour with energy.

Niida showed results on different Fourier components for identified particles in PHENIX at top RHIC energy, and about two particle correlations with flow subtracted. The results offer strong constraints on hydrodynamical models and on the jet-medium interaction.

Turning to the phenomenological side, *Werner* showed recent improvements in the EPOS model motivated by the measurements of two particle correlations - the ridge - at RHIC and the LHC. By studying regions of different energy density and for different energy losses of particles, regimes of hydrodynamical evolution, of hard particle (jet) production and of jet-medium interaction are defined. In this way, the model is able to describe a wealth of experimental data, and gives an explicit example of how complex modeling heavy-ion collisions is.

Jeon showed the results of the MUSIC event generator that solves event-by-event viscous hydrodynamics. He showed how viscosity quickly smooths the irregularities present in the event-by-event initial condition, and how shear viscosity reflects on the different Fourier coefficients.

More on the theoretical aspects, *Bhalerao* showed how important the specific realization of viscous hydrodynamics is for the description of data. He discussed how to derivate it from kinetic theory including non-local terms in the Boltzmann equation.

Finally, *Q. Wang* discussed the calculation of transport coefficients using perturbative techniques, including elastic and inelastic (radiative) processes. He showed the discrepancies existing in different calculations - some of them seeming able to describe the fast isotropization that hydrodynamical models need to describe experimental data, some of them not - and localized where the discrepancies came from.

5. Hard and electromagnetic probes

Hard probes is a hot topic in the field due to the possibility of characterizing the medium produced in heavy-ion collisions, either directly the temperature and energy density, or its transport coefficients. Both these aspects and the required benchmark measurements were discussed in this conference. The most recent theoretical ideas and phenomenological implementations were presented.

5.1. Electro-weak probes

Benhabib showed the results on Z and W production in pp and PbPb collisions at the LHC using the CMS detector. Such extraordinary results constitute a test of the validity of the Glauber model and start to offer constraints on nuclear parton densities, particularly through $W^{\pm}$ asymmetries. The extension of these studies to pPb collisions looks very promising, provided the statistics are large enough.

Campbell showed the PHENIX results on photons, both in the high transverse momentum domain where they constitute a benchmark measurement and perturbative techniques are successful, and at low transverse momentum where they show hints of thermal emission that are currently used by models to extract the initial temperatures and energy densities of the medium. She also showed the elliptic flow of photons, that at large transverse momentum is small and compatible with perturbative expectations, while at low transverse momentum is very large and defies existing theoretical explanations.

5.2. Heavy quarks and quarkonium

Kim revised the use of quarkonium as a 'thermometer' of the medium through sequential melting of resonances as suggested by lattice and potential models. Then the CMS data on suppression of higher states of the Υ family in PbPb compared to pp, strongly suggestive of such sequential melting, were shown. The results on prompt J/ψ were also presented, indicating suppression of b quarks. Finally, the new, surprising results about enhancement of $\psi(2S)$ over J/ψ at small transverse momentum and suppression at larger transverse momentum, were discussed and the uncertainties of the measurement indicated.

Bastid showed the ALICE results on heavy quarks, both from direct measurements and from semileptonic decays, and quarkonium, in several channels and at different rapidities. Charmed mesons are suppressed in accordance with current models of energy loss, while the situation for charmonium is very complex. It seems very difficult to understand present data without some amount of recombination, and the role of shadowing of nuclear parton densities has to be established before quantitative conclusions can be extracted from data.

Bona showed the already quite old results from ATLAS in PbPb collisions. On the other hand, new extremely nice results in pp were discussed, providing both tests of pQCD and spectroscopic information. One can only lament that such a wonderful detector suffers such lack of manpower.

Qiu showed the STAR results on J/ψ including v_2, which indicate negligible flow within uncertainties, and on Υ's. On heavy flavors, the evidences from a large b-decay contribution to non-photonic electrons, and the new measurements that reconcile the results from STAR and PHENIX about the charm cross section, were discussed together with the excellent possibilities offered by the Heavy Flavor Tracker for heavy flavor studies in the future.

On the theory side, *J. Wang* revised the recent progress in computations of quarkonium hadroproduction in non-relativistic QCD. He showed the advances in including different singlet and octet contributions for different states, not only for the cross sections but also for the polarization. He discussed the different strategies of the different groups to fit the available parameters. He demonstrated how the polarization puzzle remains and how it demands new data from the LHC and, eventually, the inclusion of relativistic corrections.

Bianchi discussed J/ψ polarization in pp collisions at the LHC in ALICE, indicating how it constrains different models and that none of them really works. Studies at larger transverse momentum are required in this respect, as present data are restricted to rather low values. Finally, he addressed the point of how these measurements are relevant for heavy-ion collisions as assumptions on the (unmeasured) polarization in PbPb collisions are required and they affect sizably the extracted cross sections.

Back on phenomenology, *Nahrgang* discussed the effects of medium modeling on heavy quark propagation in a QCD medium. He showed the amount of ingredients: production, medium modeling with realistic evolution, equation of state, mechanisms of energy loss,..., that have to be considered. He discussed the effect of event-by-event fluctuations in medium modeling that seem to have a modest effect, and how models that work confronted to RHIC data fail for the LHC, and vice versa.

Ferreiro discussed the importance of cold nuclear matter effects, particularly the unknown nuclear modifications of parton densities, for quarkonium hadroproduction. He showed the interplay between cold nuclear matter effects and the detailed microscopic mechanism of hadroproduction and how pPb data at the LHC can help to clarify the situation.

Casalderrey discussed a most nice, simple model for quarkonium evolution using a Schrödinger equation with different potentials. Understanding the fate of a $Q\bar{Q}$ pair in medium is crucial to characterize it. The model offers information about the interplay of the different characteristics of the medium for the different states.

Ko presented a model for quarkonium production in heavy-ion collisions, which considers production, medium evolution through hydrodynamics, different in-medium potentials for quarkonium dynamics, and regeneration. He discussed how several models successfully describe experimental data and how the amount of details required in the implementation depend on the chosen setup. Event-by-event fluctuations do not seem to have a large effect on this observable.

Finally, *Krein* revised the different aspects that must be considered in order to understand the interaction of J/ψ with nuclear matter. This issue is of interest both intrinsically for non-perturbative QCD and for the use of quarkonium suppression as a QGP signature. The different open problems, and the different experimental proposals to study it in antiproton annihilation on deuteron in the Panda experiment at FAIR, and in electroproduction in ATHENNA in JLab, were presented.

5.3. Jet quenching

On the experimental side, *Di Nezza* revised ALICE results on particle production at large transverse momentum at the LHC, both for charged particles (single inclusive distributions and back-to-back correlations) and for jets, for which ALICE offers valuable information complementary to that from ATLAS and CMS. The problems for reconstruction of the latter were discussed. While charged hadrons are in agreement with current models, jets offer intriguing features that remain to be confirmed and understood, together with the problems for their reconstruction in a large heavy-ion background.

On the phenomenological side, *Peigné* reviewed a calculation of how QCD radiation affects quarkonium production in a way that cannot be simply described by energy loss. Interferences between initial and final state radiation lie at the heart of the discussion, and the results were implemented in a model which successfully reproduces data of quarkonium absorption in pA collisions at fixed target energies and at RHIC. Data from pPb collisions at the LHC will be crucial to further clarify the relevance of these ideas.

Salgado revised the new developments on radiative energy loss of fast partons in a QCD medium. He discussed the limitations of presently used models, and reported on the ongoing work to understand QCD branching in a colored medium. This should constitute the basis for jet calculus in a medium and, as such, the basis for a rigorous implementation of multiple emissions in Monte Carlo models for QCD branching. While waiting for applications to jet physics at the LHC, this is a key aspect in order to make reliable extractions of medium parameters through hard probes.

Bluhm revised one of the aspects usually neglected in current models of radiative energy loss, namely how the presence of an absorptive medium may affect the radiation from a fast parton - a subset of the next-to-leading effects to be considered. He showed that radiation may be strongly reduced for some values of the parameters. While the effect may become parametrically large, going beyond parametric estimates is crucial for applications to phenomenology.

Arleo revised the use of γ+heavy quark production as a tool to constrain nuclear effects on parton densities in pA, and for determining medium properties in AA - with the photon coming from fragmentation and the heavy quark being affected by the colored medium. Again, pPb collisions at the LHC will be most useful to validate these studies, particularly if statistics are large enough.

6. To conclude

The conference has shown a large variety of results that illustrate the richness of the field. The pPb run [1] in January-February 2013 (the pilot run on September 13th 2012 has already resulted in 3 publications [2, 3, 4]), together with the future ion runs [5], the upgrade plans of the different experiments, and the projects for electron-ion colliders [6, 7], indicate how active the field is and evidence its bright medium term future.

Acknowledgments

I would like to thank the organizers for the invitation to such a nice meeting, and all the speakers for information on their talks and most useful discussions and comments. I gratefully acknowledge financial support by the European Research Council grant HotLHC ERC-2001-StG-279579; by Ministerio de Ciencia e Innovación of Spain grants FPA2008-01177, FPA2009-06867-E and Consolider-Ingenio 2010 CPAN CSD2007-00042; by Xunta de Galicia grant PGIDIT10PXIB 206017PR; and by FEDER.

References

[1] C. A. Salgado, J. Alvarez-Muniz, F. Arleo, N. Armesto, M. Botje, M. Cacciari, J. Campbell and C. Carli *et al.*, J. Phys. G **39** (2012) 015010 [arXiv:1105.3919 [hep-ph]].

[2] B. Abelev *et al.* [ALICE Collaboration], arXiv:1210.3615 [nucl-ex].

[3] B. Abelev *et al.* [ALICE Collaboration], arXiv:1210.4520 [nucl-ex].

[4] S. Chatrchyan *et al.* [CMS Collaboration], arXiv:1210.5482 [nucl-ex].

[5] J. M. Jowett, "LHC heavy ion programme", contribution no. 164 to the Open Symposium of the European Strategy Preparatory Group (10-12 September 2012, Krakow, Poland).

[6] The Electron Ion Collider Working Group Collaboration, C. Aidala *et al.*, *A High Luminosity, High Energy Electron Ion Collider*, `http://web.mit.edu/eicc/`; D. Boer *et al.*, arXiv:1108.1713 [nucl-th].

[7] J. L. Abelleira Fernandez *et al.* [LHeC Study Group Collaboration], J. Phys. G **39** (2012) 075001 [arXiv:1206.2913 [physics.acc-ph]].

Institute of Physics
Dirac House, Temple Back
Bristol BS1 6BE UK